DÉC
TIPS

MW01674430

MANTRI PRAGADA MARKANDEYULU, Litt.D.,
Poet, Novelist, Song and Story Writer (The Scholar)
B. Com, DBM, PGDCA, DCP,
(Visited Nairobi-Kenya, East Africa)

- ***Global Honorary Advisor, Federation of World Cultural and Arts Society (FOWCASS), Singapore.***
- **CIVIC EXCELLENCE AWARD 2022 FROM UHE, PERU**
- Rabindranath Tagore Literary Honor 2022 (Government of Seychelles, Motivational Strips and SIPAY Journal)
- **CESAR VALLEJO AWARD 2021, UHE, Peru for Literary Excellence WORLD** WRITERS' UNION Peru

- Gujarat Sahitya Academy and Motivational Strips LITERARY EXCELLENCE Honor
- *Honored with "Royal Kutai Mulawarman Peace International Institute, Philippines"*
- *Royal Success International Book of Records 2019 Honor, Hyderabad-*
- The Silver Shield Award from UHE, Peru for my Literary Excellence 2021.
- 2021 GOLDEN EAGLE WORLD AWARD FOR LITERARY EXCELLENCE, *India*
- *The Scholar, Institute of Scholars Research Excellence Award-2020, Bangalore (India)*
- *Gujarat Sahitya Academy and Motivational Strips 2020 Honor, Gujarat-India*
- *Hon. Doctorate in Literature from ITMUT, Brazil. (2019)*
- *Literary Brigadier Honor (2018) from Story Mirror, Mumbai, India*
- *Spotlight Superstar Honor (2018) from Story Mirror, Mumbai, India*
- *Golden Ambassador General for Development and Peace at @ TWPF/BTYA, Bangladesh*
- *State of Birland at Bir Tawil Recognized Poet*
- *RKMPII Nobility Award 2021*

- *RKMPII HEART OF GOLD NOBLES Honor Certificate 2021*
- *ISFFDGUN Internationally Accredited Certificate 2021.*

....

Published Books:

(1) AGENT VISNU 999 (Story and Screenplay) (2) THE BANK THIEF (3) THE TRUE HERO (4) RICH MIND (5) TUNE, SING AND DANCE POEMS (6) SOUND OF SONG (7) SUCCESS KNOW-HOW (8) LADY POLICE (9) THE CRAZY BEGGARS (10) ENTANGLEMENTS (24 STORIES) (11) POETRY LYRICS (WORLD POPULAR) (12) SPECIAL QUOTATIONS (80 Photos) (13) ANIMAL STORIES KIDS (14) THE DEAD AND GHOSTS (15) CLASS 1-8 COMPUTER LEARNING, (16) ONE STANZA POEM (17) MARK'S POEMS (18) WINGS OF LOVE (19) MY DARLING (20) 1000 PROVERS (21) NATURE & FORESTS (22) SOCIAL IMPOTENCE (23) NGO WORLD (24) SUCCESS KNOW-HOW (25) VISHAL AND THE EVIL KING (26) MANTRI'S MICRO POEMS (PART 1,

2 AND 3) (27)ONE STANZA POEMS (COLOR) (28) రావే రావే బాల (29) మోహిని (30) బావంటే బావ (31) బ్లేమర్ (32) రాణి, IPS (Telugu) (33) తెలుగు పాటలు, 100+ (34) వీర విశాల్ (35) దయ్యాల గోల (36) మినిస్టర్ ప్రేమ్ (37) ముళ్లబాట (38) చిట్టి కథల లోకం (39) బుజ్జాయికి చిట్టి కథలు (40) పసందైన బాలల కథలు (41) సామెతలు. (42) ఒక్కసారి మోహిని (43) 80 రోజుల్లో భూప్రదక్షిణ (44) సాయిసంకీర్తనామాల (పార్ట్శ 1,2,3,4,5) (45) ఆగిన గుండె (46) TELUGU SONG LYRICS AND MINI STORIES (COLOR PHOTOS)

....

DÉCOR AND INTERIOR DESIGN TIPS

INDEX

SPECIAL TIPS TO KEEP YOUR HOMEWARM IN THE WINTER

SPECIAL TIPS TO KEEP YOUR HOMEWARM IN THE WINTER

Make your home and cozy, with warm colors, textures and winter themed accessories.
To beat the winter blues, add cheer to things around you by changing your décor. Create a warm, comfortable and appealing environment to counter the gloomy weather outside.

LIVING ROOM

Arrange the seating pieces so that you can have a cozy and comfortable space. There's nothing better than a cozy corner to curl up and catch up on reading or to chat with near and dear ones. Re-arrange furniture to have some intimate spaces, and creative niches. Group some easy to move furniture's in corners for some informal conversation. Set the mood of the season by putting some dried flower arrangements or branches on your coffee or side tables. Keep cards and board games handy for family members and friends to enjoy together. This is also the perfect time to go for bright wall murals.

You should even hand a colorful painting to spice things up a bit. Place some fresh flowers in a bright and pleasant looking vase. Use of colorful and scented candles will add color and warmth to your living room.

COLOURS

The faded, grey tones outside can be quite dampening on the mood and spirits, a good way to counter this is through the inclusion of warm colors, rich textures, and winter themed accessories. So, try accenting your décor with deep red, oranges, gold and browns. Paint just one side of a wall and bright orange or yellow. You could even have some terracotta tiles on one side of the wall. Use lots of throw cushions for added comfort. They come in a wide array of shapes, sizes, and colors. Use heavy and soft fabrics for more warmth.

Velvet, satin and other soft fabrics can make things cozy and comfortable. If you have avoided using them so far, now's the time to get them.

LIGHTING

Lighting can make or break the ambience of a room. The lighting should not be too bright. Candles can be used very effectively. Decorative candles in ornate candleholders can make a wonderful centerpiece on your coffee table. Go for colourful, perfumed candles to add a sparkle to your evenings and at dinner time. Place them in front of mirrors or with groups of plants for a unique effect.

Also, opt for lampshades that art mulberry or copper colored, Soft lighting and rich, colours will make the room warm and cozy. Decorative light stands, used moderately, can help add drama and accent light to your interiors. You can drape them along door-ways, window sills, or cabinets to add a warm glow to your home.

RUGS AND CARPETS

Winter is the best time to flaunt your lovely rugs and carpets. Nothing spells winter like a heavy, furry rug under your feet. Regardless of their size, these add a sense of warmth to a room. Smaller foot mats should be provided beside the cot to avoid stepping off the bed onto a cold floor and at the entrance to the attached toilets. You can even use runners in the connecting passages and hallways

.

===

ESSENTIALS OF A HOME OFFICE

Many start-ups begin operations from home. Here are some tips to make your home office more functional, efficient and inspiring.

Ever since Bengaluru became India's start-up hub, work from home's a new trend popular among many entrepreneurs. So, home offices are here to stay. This could be a small nook in your living room, a spare room or even the space under your staircase.

It is quite important to make that separation between work and home in this pace. You may think it is fine to use your old dusty studyable, but it is essential to have a separate space for work, so you are not temped to work all the time and can shut sop at the end of the day. Some of these home office essentials will help you separate work and home, and achieve work-life balance

DESK SPACE, SEATING

Use a desk that accommodates your desktop computer and gives you space to take down notes or even doodle. Keep your desk clutter-free and devoid of distractions. A desk with an adjustable height system to suit your height and comfort level makes a huge difference. It is totally worth investing in an ergonomically-designed, comfortable chair. While choosing one, consider the long hours you will work for, the nature of work and the environment you will place it in.

TECH IT OUT

Today, nothing works without the Internet. So, make sure you have your Wi Fi set up. If you need to, invest in a good printer and scanner, and find the right location for them. Plus, a hard disk and pen drives they come in handy despite cloud storage. Also, make sure you have extra plug points.

STORAGE

Storage is pretty integral to any functional home office to store your files, reference books, stationery or even the printer. It could be a storage cabinet over your desk, a file cabinet under, or a free stand shelf next to the desk. Are you on a tight budget?

Open shelves with colorful plastic trays could work and even add some style to the room.

LIGHTING

Natural lighting is a must. It is a well-known fact that natural lighting improves productivity and work atmosphere. Find a space close to a window, preferably looking out. That can act as a gentle reminder to take a break, now burn out.

If you don't have a window nearby, use artificial lighting. A good desk lamp and some ambient lighting will help. Add some LED strips under your overhead cabinets to make the space inviting.

INSPIRATION

Pin-up boards or inspiration boards add pep to the décor and even reflect your personality. So, get creative and make your space all about you. All inspirational photographs, posters, quotes or vertical organizers lend character.

You could have regular push pin boards, exposed cork boars or even a metal sheet with magnets.

LOUNGE AREA

Even if strapped for space, try to accommodate a lounge chair or day bed for those occasional breaks. Add a small coffee table, and there, you have a lounge area to meet with clients.

SPLURGE OR SAVE

Where should you cut corners when decorating a home and what items should you spend on? Interior decorators tell you how to apportion your budget.

Whether it's time to give your home a much-needed facelift or whether you're moving into a new place and want to do it up stylishly, all your renovation dreams must fit neatly into the confines of your budget. Spend on a very fancy bar and you may be left with no money to stock it; pick up that branded recliner you've to pass on that gorgeous, designer floor lamp or painting the kitchen wall. Decisions such as these can make home renovation torturous. What items should one splurge on? Where should one cut corners?

....

START WITH YOUR GAME PLAN

Decorating a home is of course a very personal process, so your color palette and choice of furniture may be uniquely yours. However, what applies to everyone decorating a space is that she or he must prioritize. One thing to consider is what items will you get the most use out of? These, naturally, should be at the top of your priority list. If you entertain frequently, for instances, you may want to invest in comfortable bar stools, or if you love to spend your evenings with a good look, it may be worthwhile to invest in a plush reading chair and a nice floor lamp. No matter what your budget is, it's important not to lose sight of the fact that the point of this renovation project is, ultimately, to add value to your home. Whether you're opting for a complete renovation, just changing the furniture, or simply redecorating the walls, embarking on any interior designing or decorating project must be something you out your heart into. When you think of all the work that needs to get done, this can be overwhelming. But, once you have a list of priorities read, you could perhaps try to look at each point on it as a small step you need to complete, rather than thinking of all the work that will go into accomplishing the whole project. Take it one step at a time.

The thing that worries most people is whether they will be able to fit their dream décor into their budget. But, with a little direction, you can create a look that oozes luxury on a modest budget. It just boils down to good planning.

For instance, you can use accents (things like lamps, cushions, throws and rugs etc.) to perk up the ambiance in just about any room without incurring a huge cost. You'd be surprised at how dramatically these trimmings can change a room's look. A complete overhaul may not even be necessary at times if you just replace an old coffee table, get some patterned pillows and throw in a pretty rug."

WHEN TO SPLURGE

"Redesigning interiors calls for putting the focus on optimizing the space and functionality of a room. Sometimes, you've got an idea on how you'd like to decorate a space and the come across a statement piece that fits perfectly into your vision. This could be an imported lighting feature, am elaborate mirror for a wall, or even a pair of armchairs whose design cannot be easily replicated by a carpenter. Occasionally, if it's something that you really have to have, it's okay to splurge. It is also not a bad idea to spend a little more on new, energy-efficient appliances like a refrigerator or microwave. Think of these as investment pieces.

As you will spend lot of time on a sofa or on your mattress, these things need to be comfortable. There's no point in trying to save on these features, as a cheap couch or mattress could end up costing you a lot more you may develop back problems.

Designer clothes are great confidence boosters. It's true, they may make the wearer feel special, but they are also usually cut very well and so they fit better too. Similarly, a house with a well decorated interior does more than boost your happiness quotient. Spend on the right things for the right reasons and it will actually serve to boost your general wellbeing too. A good rule of thumb, he feels is to spend more on items that can affect your health: splurge on items for the living area where you spend most of your time. Good quality sofas, beds, mattresses, refrigerators and kitchen hobs will be worth every rupee you spend on them.

WHEN TO SAVE

Reserve a smaller part of the budget for the walls and ceilings. You can also save on flooring, as there are many options in this area now: you can use natural stones or a variety of tiles to adopt just about any new look that you want to achieve. Take the stucco pain finish, for instance. This is a durable option that's also pleasing to the eye. You can cut corners when it comes to wall art too. Try refurbishing an existing artwork or enhancing it with better lighting to save costs.

To this list, Accents are where you save. Don't make impulse purchases and throw away your hard-earned money on purely decorative arte-facts or on items like corner tables that you don't really use much.

Spend more on the couch, but less on the throw pillows; invest in a sturdy, stylish dining table, but buy inexpensive China in beautiful designs for daily use. And, if you simply must use that exorbitant fabric for your curtains, choose a design such as blinds that utilizes less fabric.

OLD CHESTS FOR FUNCTIONAL AND DECORATIVE USE

OLD CHESTS ARE PERFECT MULTIFUNCTIONAL FURNITURE PIECES FOR HOME INTERIORS.

One piece of furniture that has great utility and has been around since centuries is the simple storage chest. Old chest or trunks are exquisite pieces that serve both as wonderful decorative accents and as useful storage containers. They can be included anywhere in a décor as long as you know how to iterate them in a room and how make them a great focal point.

DÉCOR AND STORAGE

There are lots of ingenious and creative ways for using an old chest. In the hallway, they would make a grand entrance, and in the living room, a great conversation piece. They art multi-purpose piece of furniture and storage that serves as a decorative piece and provides lots of extra storage space right in your living space. It can be moved from home to home and from room to room, useful l I every place you put it. A beautiful, dark-stained trunk would really stand out in a bright décor with pastel, pale colors. Indian royalty used highly decorative chests for jewelers and trousseau. Even today, these chests come in numerous varieties and have different uses. It is available in all shapes and sizes; tall and narrow with lots of drawers on top of each other, or short and wide with drawers' side-by-side. It can be small enough to serve as a nightstand with drawers for necessities, or large enough to hold a whole season of clothes.

Most old trunks found today appear to have originated between 1820 and 1920 and are made of wood, with side handles, ornate carvings, and simple locks.

ROOM FOR MORE

Reminiscent of storage used centuries ago, chests have an aura of mystery surrounding them that evokes tales of travel, adventure, and days gone by. Moreover, given that an old chest is often an heirloom gift which is passed o to several generations imparts more allure to them. Hand-carved antique chests make for a great coffee table, or accent table in the living room, dining room, or bedroom. They are portable and you can take them with you when you move. They be moved from one room to another or rearranged in the room if you choose to redecorate.

You can also use it to hold a plant or to display a collection of crystals. For this purpose, you have to keep the chest open. With suitable shelving and lighting, the presentation looks attractive. For better display quotient, mount a mirror on the inside of the lid.

MULTIPLE ROLES

In the bedroom, you can place the trunk at the end of the bed where it would serve as a great place to store blankets, pillows and other similar items. A short, low chest with two drawers works well for a bedside table. Add a couple of vintage decorative pillows on the bed and you have a cozy bedroom with unique details.
In a child's room or nursery, one can use old chest to put away child's toys.

If you happen to get hold of a beautiful old treasure chest then you can even use it purely as a decorative piece. It doesn't have to be repurposed as a bedside table, coffee table, nightstand or anything else. Just use it as a place to display a few decorations and place on the hallway or somewhere where everyone can see it and admire it.

WINTER IS WHERE THE STYLE IS

You can do a lot for your home this winter by choosing the right furnishing material, color and style to blend with the theme.

You can bring in the warmth and remove the chill from your home this winter by giving your usual home décor a makeover. Yes, it is that easy all it takes is a magic wand and a whirl and swish to your local shop or favorite store.

Here are some tips to spruce up your home with winter furnishings. Some of the furnishings you can get for your home are for bathroom mats, blankets for beds of all sizes, quilts, childhood sleeping bags, room dividers, tablecloths, throw cushions, beanbags, silk and velvet bed spreads and knitted upholstery. Rugs are also trending.

COLOURS AND STYLES

The colors to look forward to are white, black, grey, charcoal, emerald, greens, browns, and kiwi hues. The styles to lookout for are block-prints, king-sized lengths and widths, fluffy, snuggly, thick and soft tassels, psychedelic artwork and vintage prints.

DESIGNS

This season is fun as well. Indian art like Worli drawings of snow-capped mountains, rain-fed rivers and earthy, rustic native designs are spread out on upholstery and dividers in some popular stores dealing with funky designs.

Mirror-work is winter-centric and fashionable. They bring the sun in especially near windows. Wreaths hanging from cushions and curtains are in fashion.

FABRICS

Living rooms swamped by linen, khadi, silk, velvet, jute, wool, corduroy, flannel, fleece, and fake fur work well. They are some of the sought-after fabrics. Faux leathers especially are being sold in assorted sizes. Blankets, cottons tweed, woolen blends woven through intricate embroidery, plastic sleeping bags and old-world silks have proved to be winter essentials. They inhere a subtle delicateness, thereby inviting chilly attention.

MIX, MATCH

While faux leather does come in black it adds an element of cleanliness when wrapped on cushions. Wooden chairs get cold during winter and leathered pillows can be placed on them. Use leather for seating.

Mix and match depending on your style but do keep in mind that juxtaposed images, prints, designs and colors are fashionable and that dark shades, though they retain heat, are gloomy and depressing. Use them sparingly because color reflects mood and can add gloom, or happiness, instantly.

TIPS

Try not to get dark with cloth as far as possible. Let all mirror-work be close to windows and where light seeps in, so that they brighten up the living room. By juxtaposing designs and retaining light through the use of bright shades, your winter home will find its soul. By maximizing the potential of your furnishings in order to bounce off light through mirrors, other shades and images will reflect off even more, hence creating a wonderful canvas of light. By being creative and in innovative, by bringing back bright colors and Worli artwork, old-world Indian grace is back this winter. The best part of this year's winter's furnishings and furnishings and fabrics is that they can be soft and rough. While velvet duvet can put your children to soundless sleep, rough jute and khadi on chairs will keep them up studying till night. This season's mattresses ought to be covered with quilts instead of thin bed spreads. Innovate by using whatever is present in your drawing and bed rooms and the result will be a snazzy, less breezy and spacious interior that hinges on the opulent.

MODERATE HOME MAKEOVER

Small changes can make a huge difference to your home.

If you are a little bored with your living space, it might be time to spruce up your apartment.

FEATURE A WALL TO STAND OUT

While walls are safe but a little boring however, painting all the walls in a room with non-neutral colors can make it look garish and worse, create sense of claustrophobia Designating a side as a feature wall will allow the color to stand out even morel You can either apply a coat of texture paint or paste wallpaper. While texture pains have a richer look, wallpaper has the advantage of being scraped off and easily replaced. Today, texture paints arrive with a lot of creative embossed designs while wallpaper patterns run the gamut from dignified too cheeky.

BLIND UP YOUR WINDOWS

Cloth curtains are passé and wooden blinds are in vogue. Nott only do they class up your windows, they can also regulate the amount of light you want to let in. They are also easier to take care of and unlike curtains, don't require to be changed year after year. For those too attached to curtains, opt for heavy cloth curtains at the ends of the window rod and light sheer curtains in between. Sheer curtains can be used during the day while cloth curtains can block out your windows during nighttime.

INTRODUCE A WATER BODY

According to vastu, a water body in the north east of a room harness positive energy. Filling up a brass uruli with wager and strategically placing it in the living room will infuse an element of serenity into your home. Rose petals add a nice touch. If you are willing to spend a little more, you can opt for a stylish contemporary lamp with running water, which manages to be new age while evoking nature.

MAKE FOR PARTITION SPACES

It is a good idea to demarcate large connected spaces especially when the living room, dining area and kitchen are all connected. A top wooden panel supported by Chettinad pillars is perfect for someone who wishes to add a regal and ethnic touch toll their home. For those who prefer contemporary designs, sliding glass doors could do the trick.

THE RIGHT KIND OF FLOORING

Retiling can also dramatically change the way your home looks. Light colored tiles increase the sense of space and are especially apt for both rooms. For the shower area, pebble tiles can give a stylistic yet calming effect of white pebbles on the floor. Wooden flooring in the master bedroom or the entertainment room can give the place a more restrained look. However, restrict its use to areas that are not water prone.

CARVE OUT A NICHE IN THE WALL

Niches in the walls also add pizzazz to your interior décor. Being clear on the right size of your niche is important. Well-paced niches, especially those with lights that use concealed wiring are perfect places to display your curios. They are also ideal for placing your AC stabilizers, so that they don't jut out in your living room Cover them up with louvred doors to give a neat look.

PERK UP DÉCOR WITH PEBBLES AND STONES

Enliven your garden and outdoors with interesting accents like stone cladding and pebbles.

Our home needs a revamp once in a while to keep up with the interesting trends in décor and architecture. However, space cramped we may be, it takes a little effort ad time to do up your garden and outdoors. Landscaping has evolved so much over the years that it is not an outdoor thing anymore. One can do up corners and spaces inside the home too.

This natural material lends texture and a rustic appeal in the interiors whether used in its polished or unpolished state.

A natural material that works wonders as a décor element is stone. It is not only attractive but also durable. The varieties of colors and textures in stone make it an interesting element of décor in homes. Apart from its use at the construction stage, stone as an accent looks extremely lavish in beautifying the home interiors. This hard and durable material can be cut and sculpted into fluid forms as an element of décor.

BEAUTY OF STONE

Generally, stones such as marble are graded for soundness for appropriate use as a building material. Slate, limestone and vernacular stones such as Cuddapah or Shahabad are beautiful and contribute to the character of the interiors. The beauty of natural stone is that the color is not uniform or typical. What may be a flaw or vein of impurity in the stone is also the reason for its color or uniqueness.

STONE THEME

You can opt for a stone theme in your home décor wherein it can be used extensively in the entire house. Stone tiles, decorative stone walls, archcs and columns can be suitably tailored to be contemporary or traditional.

Stone can be sliced into slabs which is widely used in flooring, the most common being granite and marble for cladding since it gives a more rustic look as compared to a polished glossy surface. Granite is a form of igneous rock and is ideal because of its hardness and non-porous nature. Granite comes in wide variety of textured finishes called antique finish, leather finish, etc. where it is rendered rough by first flaming and again brushed with a polishing stone for the desired effect. This finish is widely used where an anti-skid surface is required.

You can choose stone floors with natural polish as flooring. All these give the floor a completely rustic look. Stones require less maintenance and give warmth as well as beauty with a completely natural look.

Stone can also be used as cladding in the forms of slabs and tiles. There is a wide variety of slate stone, limestone and sandstone available.
A dining table with a stone top is perfect for a rustic ambience. With colorful inlay work, it can become the highlight of the dining room. Simple blocks of stone can be used as

seating, display stands. Or you can use slabs of stone as shelves, decorative ledges or as table tops with wood serving as the base.

USE IN THE OUTDOORS.

Stone is widely used in landscaping with its ability to transform the outdoor ambience and because of its rugged ability to withstand all kinds of climatic conditions. For the portico, if the decor theme is contemporary, the flooring could be of stone, with the walls featuring stone cladding.

DÉCOR WITH PEBBES

Pebbles are also a popular trend in stone-themed home décor. It gives antique look and creates a peaceful aura in the room. Decorative pieces designed with pebbles are easily available. Pebbles are used not only for interiors but also in gardens and fountains. They can also be layered on an open porch or in verandas. The soothing colors of stones and the gentle rounded appearances evoke a natural feel to the house.

The variety of colors and textures in pebbles make them an interesting element of décor in homes. They are available as natural and artificial stones, either smooth or rough finished. The look of the stones depends on the material it is fashioned from such as marble, granite, quartz or ordinary ones from riverbeds.

PLACING PEBBLES

Before replacing an area of grass with gravel or pebbles, decide which kind of pebbles will suit your space. Pebbles vary in size and can be obtained depending on their use and locating. The smaller ones are excellent add-ons to other aspects of decor.

They can complement a dry flower arrangement, enhance the base of a bonsai plant, or complement a fresh bouquet. Smooth pebbles are used in aquariums as well as to line indoor water bodies and fountains.

Also, pebble tiles are definitely becoming more and more popular. Designers are using pebble themed floor tiles to add punch and definition to floor areas. Decorative items made from pebbles are also in vogue these days. These are easily available and at prices which are quite affordable.

They are also used as tabletop decorative piece. A bowlful of pebbles on the central table can give trendy look. Stones and pebbles can also be used as an eye-catching wash basin.

FLAUNT YOUR KITCHEN IN STYLE

From modular kitchen boutiques to designer kitchen, the retail market today is set to woo its customers with everything you need to transform your kitchen space into a fabulous one.

TECGHNOLOGY INTEGRATION IN HOME FURNITURE IS A TREND TODAY AND AN INTERESTING EXAMPLE OF THIS IS THE TRENDY MODULAR KITCHEN.

Homes are a reflection of one's own self as it reflects the taste, style and personality of its owner. Design and aesthetics, therefore, play a key role in the décor of a home. Technology integration in home furniture is a trend today and an interesting example of this is the trendy modular kitchen.

Modular kitchens seem to be moving in good volumes as it is user-friendly and comfortable. Traditionally, Indian hoes look for functionality. Hence, today we see a combination of functionality and aesthetic playing a key role, and products which have the combination of these, move faster.

Trending ideas in modular kitchens include an open or island kitchen, advanced technologies such as automated handle-free drawers and lighting devices like glass units with internal lights. Besides the open and island kitchens, different types of kitchens such as the traditional style kitchen, cottage style kitchen which is an informal space, urban style kitchen that's uber cool and chic, and Mediterranean style kitchen, which is essentially about colors.

When it comes to colors in modular kitchen, people prefer bright color combinations, such as black and red, green and white, grey and silver, etc. Moreover, choosing the right material for modular kitchen is very important as it decides the life of the space.

Kitchen cabinets have transformed its look and functionality over the years to suit the chic designs in the kitchen. For kitchen cabinets one should opt for hydraulic doors and drawers which are easy to operate and roll out. Cabinets come with foldable doors that close at the touch of a button. Shaker cabinets and those with horizontal orientation are also gaining ground.

DREAM KITCHENS

Our kitchens reflect our personality and therefore, should be less crowded or more inspiring But with the number of appliances and cutlery we use, this is difficult Smart storage space is the key here.

The trends in kitchen design include innovations on the cooking front too. Imported kitchen accessories such as air fryer electric grill, 3 in 1 gas oven tandoor, built-in gas burner with hood, ceramic hot plate, and electric barbeque and so on makes it more appealing

Finally, remember a kitchen is all about comfort as well. A home décor expert can only work on the walls of the kitchen. It is the owner of the kitchen who can make it look warm and inviting. This is why modular kitchens are here to stay.

GLAM UP YUR KITCHEN

HERE ARE A FEW HANDY TIPS ON HOW TO SPRUCE UP YOUR KITCHEN AND GIVE IT A COMPLETE MAKEOVER.

The kitchen is one of the most interesting and happening place in any house. But somehow, it remains underrated and lacks cleanliness most of the time Check out these tips to transform this lively space into a happening one:

SPRUCE UP YOUR WALLS

Splash your walls with a bright color (mango orange, sky blue, lemon yellow). Another personal favorite for me is to use blackboard paint for the wall in your kitchen that has the most accessible blank space. You can use this to scribble your recipes, shopping lists, to do's or just fun food sketches. It adds a special personal touch!

GROW YOUR OWN HERBS

Add in a small grill outside your kitchen window and plant your own herbs. There is nothing as satisfying as topping dishes with freshly picked herbs. The burst of flavor and vibrancy of color will take your food up so many notches and it's all organic. Some of the plants that are super-easy to grow are mint, basil, lemongrass, chilies, karipatta, etc.

USE ALL YOUR WALL SPACE

Many of us face a space constraint in our kitchens. I would recommend using any wall space sensibly. A great way is to put in some hooks and hang things that are non-stackable, for example, pans mugs, etc. They add a homely touch to your kitchen and they save you that extra space in your cabinets. In case wall space is limited, you could suspend a rod from the ceiling and use it for hanging too Fit in hooks and handles to the exteriors of cabinet doors to hang your kitchen dusters and gloves here.

ERASE THE DINNER-SET MENTALITY

Go eclectic on your crockery.
Choose a palette earthy or pastels or bright and buy your crockery accordingly This way you can keep adding to your collection each time you see something that fits in instead of putting all your eggs in an expensive set that you will get bored with soon anyway.

RECIPE FOLDER

Many of us have loads of cookbooks or recipes downloaded from the internet but let's face it – we really use or would like to reuse only some of these. Create your own book recipe folder with printouts or handwritten pages of all your favorites. Organize these by cuisine and handwrite your notes and changes so that they are all recorded. Keep this handy with you in the kitchen you will not waste time searching for that recipe you made a year ago and feel like revisiting today, and you won't be stuck when you can't recall the tweak you made to it that made it just perfect for you.

KEEP A WELL-STOCKED LARDER

Often the enthusiasm to cook something is wiped away because we don't have the ingredients. Place permitting keep a basic stock of oils, sauces, pasta, spices, grains etc., so that you are ready to go. Keep these in transparent containers to you can see what you have stored and tack on the expiry dates so you know what to use before it spoils.

FIVE ELEMENTS OF FENG SHUI

FIRE

It consists of colors like red and orange, and materials like electricity. The element is determined by the shape of a triangle and by objects like candles, fire place, stove, television, stereo, lights, electronics, etc.

EARTH

It consists of colors like brown and yellow, and materials like brick, stone and ceramic. The element is determined by the shape of a square and by things like masonry, marble, tile, pots, etc.

WOOD

It consists of green and blue colors, and materials like wood and paper. The element is determined by the shape of a rectangle and by objects like plants, wood items and books, etc.

METAL

It consists of colors like white and grey and metallic objects. The element is determined by oval and round shapes and by objects like pots, pans, silverware, appliances metal items etc.

WATER

It consists of colors like black and dark blue, and materials like glass. The element is determined by wavy and irregular shapes and by objects like sink, tub toilet, faucet, aquarium, pool, glass items, plumbing etc.

ACCESSORISE YOUR FLOORS WITH FLAVOUR

Spread the cheer with a new décor on the floor in your living spaces.

Spring is here and it is time to usher I the New Year with a new look for your home. As you deck your homes with thorns of tender mango leaves and rangoli, add some festive flair to your floors. Fresh green, orange, bronze, gold and pink home décor accents are the trend this season.

Explore and experiment with new décor ideas that range from the traditional to the offbeat.

STARS ON THE FLOOR

Bring the stars down to light up your home differently for all Festivals. Clay diyas and dainty wax ones look ethereal floating in an uruli of water. Their twinkling flames reflecting off the water cast a spellbinding aura, spreading the festive cheer. Place a few uruli with these floating lamps in the middle of each room.

Explore the exotic. Try an Oriental theme for a change. Place foldable Chinese lanterns or delicate looking sky lanterns around the living room. Sky lanterns are made of rice paper and bamboo casing with a candle in it. Once the candle is lit, the heat causes the lantern to rise and float. To place the sky lanterns on the floor, make a small opening at the top so the heat escapes through this vent and it does not rise.

MEDALLIONS ADD MAGIC

Create an element of surprise and tread a different décor trail for the occasion. Install a easy made floor medallion with intricate patterns and motifs beautifully crafted in marble, inlay work and mosaic. Mosaic medallions are composed entirely of tiny marble tiles that can be worked more easily into curves and delicate patterns than can large pieces of stone. These medallions come in one piece mounted on mesh backing, and are available in a variety of sizes and colors.

They can be easily installed with minimal effort and require only precise grouting to round off the task. These are long-lasting and can take on the role of an everlasting rangoli.

MIRROR MYSTIQUE

Mirrors in all shapes and sizes can reflect the festive mood in more ways than one. Try and have them around the house in various forms especially on the floor as an art form. Mirror work embroidery on throw pillows and lined on durries make wonderful upholstery displays. Spread these on the floor near the urulis or in doorways.

Clear up an entire wall to display a cluster of mirrors in all sizes with ornamental frames. Use a couple of hand mirrors creatively by lighting candles on the handles. Place these on low stools in each corner of the rooms. The flame bouncing off the mirror can cast a mesmerizing halo around the home.

FLORAL FLAIR

Keep traditions alive with a typical festive welcome at the doorway. String along festoons of marigold and mango leaves across the door top and let them hang delicately on the sides. Instead of one string, bunch a cluster of them so you have three or five garlands of these on either side of the door frame. Extend these across the doorsteps into a encompassing floral theme.

Experiment with a medley of all the seasonal flowers for a vivid burst of happy colors. Don't ignore the inner doorways of your home. Go in for different flowers to grace each inner doorway. Spread the natural floral fragrance with aromatic flowers like the jasmine and rose but don't go overboard as the fragrance could be overpowering.

If you have large urns, place them flanking the entrance. Slip in long stemmed fresh flowers like lilies and cannas into them.

SPLASH OF ART FOR THE FESTIVE SEASON

Artistic touches to the décor scheme with unusual artifacts and exclusive hand-painted utilitarian objects enhance the festive ambience in the living areas at home.

Festivals are a great time for families and friends to gather and celebrate the occasion. With festivals around the corner, this is the start of the festive season, which calls for a special décor to emphasize the holiday spirit and to celebrate it with colours, creativity and renovated interiors. Integrating artistic trends in interiors is a wonderful way to express creativity and fashion unique, individualized décor themes.

Artistic expressions need not be limited to wall paintings but can also be functional and creative, and extended to utilitarian objects and artifacts. Utilitarian objects from around the house can be the perfect canvas for painting and are often transformed by artists into works of art. Wood cabinets, chairs and tables shelves, dressers, boxes and chests when hand-painted make great statement pieces. These can be in traditional designs, use religious motifs, have a modern pop art look, or even have geometric patterns and delicate detailing for an interesting effect.

Most importantly, these are usually hand-painted and available as single, one-of-a-kind pieces Therefore, these objects of art have a special attraction and lift the most mundane interiors into artistic environs. At the same time, as these are not available off the shelf, there is an exclusivity to them which helps in converting them into statement pieces.

Any interior décor can be modified to adopt various décor themes by placing artworks, artifacts and furniture that reiterate specific thematic concepts. For instance, a retro, art deco, traditional, Asian, eclectic or any other décor theme can be accentuated with the right design, form and motifs highlighted in the art accessories. The result is not only updated interiors but is also refreshing for the entire family as it gathers for the festivities

Whimsical, narrative or popular icons, any thematic subject can be painted on the surface of most materials. In general, these are commissioned to an artist when there is a specific requirement built around an interior décor, and colors and motifs have to be coordinated. Sometimes these are available as part of specifically conceptualized art exhibitions and are for sale.

These are also available as limited-edition products, where popular paintings by well-known artists are transferred onto various objects of home décor, including fabrics, cushion covers and other accessories Exclusive table mats, coasters, mugs, plates and other lifestyle products are printed with popular works by artists and are perfect for adding a touch of art to home interiors.

A splash of vibrant colors is perfect for the festive season and will add interest to all interiors. However, hand painted furniture and collectibles are also available in minimal designs and a monochrome palette, and these can add sophistication to modern and contemporary interiors as well.

MAKE-AT-HOME FESTIVE DÉCOR

Give your home a festive makeover with these simple

While most of us anxiously await the festivities, they sure come with extravagant expenditure on items of décor that are not used the rest of the year. Here are some ways in which you can use easily available items to make some festive essentials which can be re-used throughout the year.

WALL HANGINGS AND TORANS

Torans are an essential part of festive décor. While most of the traditional designs have mango leaves strung together and hung up at the entrances, the market is filled with many more designs that are made out of glossy paper, thermacol discs, etc. How about you ditch these and make your own? Go ahead and either use real flowers, or string some artificial flowers together and you have your custom-made toran. You can also use redundant CDs, cardboard, etc. to make wall adornments that are unique and tailored to your home décor.

LIGHTS AND LAMPS

No festival is complete without lights. From sea shells to that chipped cup that is unfit for use, or those mismatched glasses that you can't seem to throw away, all of these can be put to use to make some home-made candles to create a warm glow in your puja room. String some fairy lights with some flowers to add a beautiful aura to your home. Use old glass jars or bottles to create temporary lamps by just throwing in some fairy lights could it get any simpler than this?

UPHOLSTERED AND UPBEAT

Why splurge on those elaborately done cushions which you barely use throughout the year when you can make some with minimal effort. All you need are some rhinestones, craft glue and some ornate laces and you have yourself pretty and festive cushions that will add a hint of glamour to your home. Drape that silk dupatta you barely use, or that ancient saree with sari embroidery, over the sofa handle and volla your seating is festival ready.

While we have just mentioned a few ideas that can give your home a complete makeover with minimal effort, put on your thinking cap and you have yourself the unending and exciting world of DIYs that you could try to make your home look more festive this time around.

FESTIVE FLOWER DÉCOR

This Dussehra, create simple, stunning flower decorations for your home.

The festive season has begun and what better way to celebrate than to redecorate your home in the traditional Indian style. Dasara is a great festive occasion to style your living space with nature's bounty of flowers. You can't imagine Dasara without bright yellow orange colored marigold flowers. Traditionally, Dasara decorations display marigold flowers, a symbol of energy and beauty. People decorate entrances of their homes with torans made up of marigold flowers and mango leaves.

The fragrance of fresh flowers adds warmth and glow to a room, and charm to the décor. Decorating with flowers is an art and it can be developed with a little interest and effort. There's a wide variety of flora for you to choose from, which can be combined beautifully for decoration.

CURTAINS OF FLOWERS

It is interesting to have long curtains of flowers grace your living space it adds color to your space as well makes your festive décor stand out. Flower curtains are one of the traditional methods used to decorate the house during festive season especially if you are throwing a party at home.

It is an inexpensive as well as colorful way to do up your home. Just tie a long thread to a needle and inject single flower inside it till the thread fills. Finish by tying a knot so that the flowers do not fall. Leave a little thread so that you can tie it to the curtain rod without much difficulty. You can either use a single flower or blend some colorful ones just to make an attractive flower curtain. You can tie these curtains in Pooja room or all over the house to cover the windows.

FLORAL FOCAL POINT

Table top decoration forms a vital part of festive décor as flowers are as much an integral part of the festivities as are lights and colors. While exquisite floral arrangement would give a stunning look to the table making it the main focal point of the room, it is a good idea to balance it with an equal number of eateries and diyas.

Opting for a limited number of flowers but enhancing the décor with candles and exotic crystal ware is a great idea as the crystal ware along with the candles gives a rich aura while not making it appear heavy. If a simple table décor is desired, an attractive crystal bowl with flowers, surrounded by diyas would suffice. Pebbles too could be added to the flowers in the bowl to lend a unique touch.

FLOWER PETALS SHOWPIECE

A bunch of flowers placed in appropriate vessels makes the room more attractive. Decorate the doorway with flower garlands and place loose fragrant flowers in urlis filled with water.

Choose a large fancy container like a large shell, an earthy trough or else a decorative urli. Arrange the base with chicken wire and fit in sprigs of long-stemmed flowers and green fronds. Get the special feel to your living room by placing it in the center of your living room or on your doorway entrance. Spread flower petals that are bright and big. You can also light scented candles to spread its aroma in the room.

CORNER TABLES

A corner table flower arrangement looks great with long-stemmed flowers. With long-stemmed flowers, vases certainly help to add cheer to the space. Make sure that the colours chosen either match or are of the same hue.

When it comes to display, a long wide vase can hold a bunch of blooms, preferably limited to one or two varieties. This setting creates a beautiful soft effort and is ideal for an intimate gathering. And it's also a practical solution when there's a dearth of flowers. You can also have a cluster of two or three containers of varying sizes to add an impact to the look.

RIOT OF COLOURS WITH A RANGOLI

Give your imagination free rein this festive season with innovative rangolis

Festivals give us the freedom to celebrate in our own personal ways. Decorating our homes is one of the many ways we can usher in the positive energy and create a welcoming ambience. The colorful rangoli at the doorstep creates a friendly, inviting vide to welcome guests during this festive season.

FLOWERS GALORE

Traditional rangoli designs involve the dots and weaves format mostly in rice power or chalk sticks. These moved on to the use of color powders for an artistic appeal. Experiment with some flower powder this festival season. The city is known for its flurry of seasonal local flowers, so choose which flowers will last long and not wilt after a couple of days. Choose from the eternal favorites the rose in all its hues, jasmine, marigold, hibiscus all colors, chrysanthemum, balsam, as well as the lily and lotus. You could remove the petals or use the whole flower this depends on the design you have planned. Pluck some leaves of different sizes as well as some grass.
Draw a pattern on the floor with chalk and work around the outline first. Dark colored leaves will do the trick for the border. Depending on the design, deftly fill in the spaces within the borders to you have a beautiful carpet of flowers at your entrance.

Place clay or brass lamps within the flowery area. When these are lit up in the evening the glow from the lamps falling on the flowers can create an ethereal appeal. The natural fragrance emitting from the rangoli can soothe the senses and pervade the festive atmosphere.

SAND STORY

Colored sand and multihued pebbles are another innovative way to do up your rangoli. Remember, the different colored sand once mixed cannot be separated. So, plan the design in such a way that you have different demarcated spaces for each amount of sand.

Once you've drawn the design, the marcate it with pebbles for the corner. Use different colored pebbles for each detail. For instances, if your rangoli design is about flowers in a vase, use large black pebbles for the vase outline and smaller lighter stones for the flower borders.

Once this is completed, pour and spread the colored sand within each frame so the sand is not mixed and you can use it all again till the next festival.

Innovative with designs according to the festive theme or go off beat and contemporary for a change.

MIRROR MAGIC

Give vent to your imagination this festive time. Fashion out a neat rangoli design and work your way around tiny mirrors and colored glass beads. This could be a table-top rangoli to fit onto your coffee table or the side tables. Plan-out a large rangoli for the coffee table and smaller ones on the side tables to complement the focal rangoli design.

Once you have figured out the design, arrange the small mirrors along the outline and fill in the spaces with colored glass beads or even colored sequins. Place tiny silver lamps or figurines in the center of the rangoli to draw attention to the design. When lit up, the mirrors will reflect the light from the lamps and cast radiance all around, with the glow from the side tables adding to the effect. Festive times are here. Unwind and give free rein to your creativity with rangolis.

SOME QUICK DÉCOR TOUCHES

Rangoli, a popular traditional floor decoration, is a centuries-old art form which is associated with festivities and auspicious celebrations at home.

Major renovations and redecorations in homes are planned well ahead of festivals so that the interiors are ready for the final makeovers during festival days. This gives you time to add to finishing touches to the interiors in terms of colors, accessories and traditional objects which are meaningful and also enhance the décor. You can also rearrange art and artifacts for maximum impact.

Rangoli is a popular traditional floor decoration which is associated with festivities and has its origin from ancient times. Considered to be auspicious, rangoli is conventionally drawn by women with their hands and is typically made in front of main doors of homes and places of worships and is typically made in front of main doors of homes and places of worship as a sign of welcome Rangoli patterns are made every day in several parts of the country. It is an essential component of most homes during festivals. The patterns in white, or rich colours and flowers, is an expression of creativity and dramatically alters the décor.

During Diwali rangoli is made in many parts of the country and is considered to be a welcoming sign for Goddess Lakshmi. Lotus, the symbol of Goddess Lakshmi, is a commonly-used motif. However, you can create any rangoli design or follow traditional patterns based on various regions. These could be geometric in shape, have a ventral motif and move outwards with outlines, or you can trace elaborately detailed pictures and symbols with rice flour, colored sand and flower petals. You can also use lentils; rice paste and natural colours such as turmeric and kum kum – colors that are considered auspicious. White is also used popularly as a symbol of purity. You can also coordinate the rangoli colours with the predominant colours schemes in the room to make it cohesive.

Along with the entrance, thc rangoli can be drawn in other parts of the house too as a way to add color and texture. Smaller patterns in combination with a large central motif are a good way to expand a theme and fill up vacant spaces. There are many variations in patterns that you can adopt, but the unifying thread is the creativity and dedication that goes on into making these elaborate and intricate designs. As a décor element, a rangoli is quick to alter the ambiance and add a festive touch.

Add torans to emphasize festivities and to make the entrance colourful. Ranging from simple ones to vibrantly colourful and elaborate torans, you can display these at the main door and puja room door as a sign of welcome and to symbolize an auspicious beginning. Religious and mythological motifs are also used in torans to create charming pieces that are creatively designed.

These last-minute additions to the home décor are easy and quick to make, and have traditional connotations as well.

BRIGHTEN UP YOUR WALL SPACE

A wall clock can be a perfect décor accessory to perk up the living areas aesthetically.

Wall clocks are the perfect accent element for your living room. It can become the focus of a wall when chosen according to scale and proportion. Clocks can even complement your personality, the colours that you like, and in general, add style and elegance to a wall or room – if you look around and observe the various different kinds of clocks, antique and contemporary, ornate or simple with minimal lines you'd find that there's a clock that matches almost any décor scheme and room of your house.

SPECIAL FEATURES

Apart from clocks available for the living room, bedroom, and children's room, there are special ones for kitchens and bathrooms too. There is an array of clocks to choose from, like, pendulum, digital, classic, cuckoo, children's and outdoor wall clocks. Contemporary wall clocks give warmth to your home décor. Besides conventional black, white and brown colours, there are flamboyant colours like red, blue, orange, green and yellow which can bring walls to life. Present in both Roman and digital numbers, it proves to be pretty formal at times.

An antique wall clock not only has great monetary value but also makes for wall decoration and the best specimens are actually each a great piece of art. There are different materials to choose from like teakwood, rosewood, oak, different kinds of metal, glass, and plastic. Many antique wall clocks are pendulum operated and some are set in picture frames. A spacious drawing room which reflects old-world elegance is the perfect place for that large, wood, classic-style clock or an antique specimen.

DESIGNS AND HUES

Of late, even eclectic art deco with motifs of the sun, moon, stars, and other new age designs are becoming popular.

Geometric designs are also popular with some. Modern-art inspired designs and abstract forms are also popular with those who like a distinctly different look to their wall clocks. There are lots of pieces which come in bright colours and are ideal for children's rooms or even the family room.

Since wall-clocks are generally inexpensive, you could actually have a collection of them. One of the ways to decorate with wall-clocks is to hand several similar ones in a row, as if they were an art gallery.
A cuckoo clock is among the most unusual type of clocks – one of the oldest types of wall clocks are the cuckoo clocks and their timelessness will never go out of style. Most of these clocks are made either in the shape of a bird-house or chalet. They have a small mechanical cuckoo bird pop out of a door on the hour to accompany the chirping noises.

Modern cuckoo clocks have a variety of music and bird songs and above all, run on quartz movements instead of old-fashioned mechanical movement.

ART IN FESTIVE DÉCOR

Choosing and displaying the right form of art makes a difference to the ambience.

During festivals, the décor nuances as some great importance, and every attempt is made to create exclusive and individual spaces which are celebratory in nature and reflect festive themes. Elements such as colours, design themes and styles are explored in order to achieve a bright and cheerful look, which is also cohesive and highlights the festive spirit.

Art is always a great addition to the interiors and can significantly alter the entire look and feel of its environs. Various genres of art, size, medium, style and period affect the environment and the ambience in diverse and distinct ways. Hence, the choice of art as an accessory in the interiors is an important consideration to complement the space and to blend in with the overall desired look.

In general, you would buy landscapes, portraits or abstract artworks to display at home. Representational art which is identifiable is often more popular than non-representational art which tends to be abstract. During festivals, traditional and ethnic décor can be augmented with art and artifacts that emphasize the festive ambience. Similarly, ethnic colours reflected in artworks accentuate the thematic setting.

Vermillion, ochre and turmeric are some of the colours that have a rich cultural and traditional association, and are ideal for a festive environment. Apart from furnishings these can also be supplemented with art that is aesthetic and rich in these colours. Conventionally, many folk and tribal art forms made use of these vibrant hues and employed natural pigments to paint on walls and later on also on paper. Today, the pigments may not be derived from natural sources but these colours are still popularly used in many traditional art forms.

During festivities, folk and tribal artworks in the form of paintings or murals can be used in décor to give indigenous art forms a boost and to add individuality to the décor. The versatility of these folk arts makes them easy to adapt to a variety of living spaces, right from the living room, dining space to the puja room. And, opting for a mural on the wall makes it easy to customize it, while exploring specific themes related to religious beliefs, mythology and rituals.

Another advantage with these is that every few years, new these and narratives can be painted on the walls. Folk art murals are also affordable and prices depend on the artist and the surface area.

You can also convert small furniture pieces such as tables, chairs and chest of drawers into unusual artifacts which are painted with religious text or depict content from mythology.

Abstract art which deploys play of colours is more neutral in content and can form a vividly colourful backdrop for the interiors. There are several ways in which contemporary art also explores traditional and folk themes, and these can be exhibited in homes to high-light the festive spirit.

DIWALI TO USHER GOOD TIDINGS

With this year's festivity still in the air, realtors in the city have already planned the upcoming year to get the best of the booming real estate scene in the city.

Diwali is at our doorstep and apart from sweets and crackers, the plans for the new financial year is also on our minds. For realtors in the city the plans for expansion in the real estate market are in full swing.

The market has picked up a great pace. If things continue the way they, are we can definitely expect a 3-4% growth in the real estate market compared to that of last year.

The city is likely to witness a number of new projects coming and the emergence of new districts will be an added advantage for expansion of the real estate market of the city. With the ways things are now, commercial real estate is definitely on the cards. The market has good potential for returns.

Apart from commercial real estate, another steady trend that has started and is likely to continue in the next year is the emergence of new real estate hot spots. The areas in the outskirts have shown great promise in the market, especially for villas and independent houses. The development of real estate in city is getting lopsided with developments concentrating mostly I around the western corner. In upcoming areas little far from the main city will show a lot of potential.

Taking all the trends into account it can be safely predicted that after years of struggling with a rather slow market, real estate scene in the city has picked up a great pace, and an all-round development in infrastructure. And, the realtors in the city are leaving no stones unturned to harness the best out of this surging market.

A GLASS APART

Glass bottles can be recycled and used as stunning decorative pieces, indoors as well as outdoors.

There is much that can be done to brighten up your home with not much apart from a little creativity. Of the many things that are usually available in a home and can be recycled and used for décor purposes, glass bottles are the most versatile. They can brighten up a space with hardly any expense. If one looks carefully, there are bottles of wine, perfumes, jams, coffee, pickle and so on, which can easily be reused creatively with imagination. Instead of chucking the used glass bottles and jars, you can recycle them. You could either opt for a DIY project or for the more elaborate projects, you could take a little help from a neighboring glass store.

- Reuse glass bottles and jars to create colourful utility and decorative items for the home.

- Either cut the bottle neck yourself or get it done from a glass cuter. Always make the edges smooth.

- Depending on your creativity and imagination, you can paint the bottle, do decoupage or decorate bottles with sequins, glitter and paint.

One can make vases, candle stands, glasses, lamps, terrarium, out of used bottles and much more

One can make lovely tea light holders, vases, pen stands or trinket holders. In fact, a bottle as a base for a bedside lampshade looks simply amazing,

Decorating the bottles can be done in various ways like spray painting, or stenciling designs onto them. One can add photos, messages, dried flowers or stick embellishments like lace, satin ribbon, glitter, beads, colored threads and sequins. Paint them or mosaic them with broken pieces of tiles and grout or simply decoupage the bottle and jars yourself a home and add a sparkle to your interiors.

The most simple and beautiful way to decorate a bottle is to decoupage it. Decoupage is the art of decorating a surface with paper cut-outs, using a various (or glue) to cover the surface. Colored papers or old magazines pages, good quality glue and a flat brush, are all that one needs to create a stunning bottle which can be used as a décor item.

Mosaic with broken tiles or colored bottles and grout that is readily available at the hardware store, is also a great DIY project. One can make a set of drinking glasses by cutting used bottles of same size into half or in whatever height you prefer. Remember to polish the edges to make them super smooth. Just grinding the edges won't help.

Take them to a glass shop to get polished for a better finish.

Before reusing a bottle, soak it in lukewarm water to peel off the labels. Glass is a versatile medium and it's fun to recycle it into decorative items.

To experiment at home with glass bottles, one needs a glass cutter, sand paper and lot of patience. At certain levels where you want t cut, make a mark with a glass cutter then pour hot water inside at the level little above the ark and from outside pour cold water. Continue the process till you hear a crackling sound and glass is cut accurately.

Cut the neck f a bottle and drill a hole at the bottom of the bottle. To drill a hole, use a drilling machine or get assistance from a hardware shop. Insert string lights and use an old wire hanger to spread the lights evenly in the bottle. When you turn the lights on, you will have a stunning lamp created at home.

Recycled bottles add sparkle to interiors – be it the living room, garden, or the bar area. As money plant holders, one can keep them on the table or hang them on windows or create a terrarium. Ideal bottles for this are wide mouthed ones and ones with a lid. A terrarium is for plants that like humidity and are small like ferns and moss. It is easy to add warmth and a sparkle to your home by recycling glass bottles with just some basic craft skills.

GET CREATIVE WITH MIRRORS TO JAZZ UP YOUR HOME

Apart from telling you that you look beautiful every day, mirrors can be used creatively at strategic points in the house to enhance the décor. Here are some ways to use them in a creative manner.

Add sparkle to your cooking: Mirrors in the kitchen can attract grease and grime, so use them in a smart way. One way of doing this is by using an antique mirror with a smoky effect. Place them in such a way that they don't attracts unnecessary soot.

Mirrors for outdoor areas: If you have an outdoor area, terrace or a big balcony you can use a clear mirror to make the space look much larger. Use mirrors with appropriate frames as per the theme of the outdoor area.

Smart use of mirrors in the bedroom: According to vastu and feng shui experts, mirrors placed in from of your sleeping area are a strict no, but with proper advice, you can place them in areas other than these. Place them in such a way that they reflect light from the lampshade.

Make the bathroom look elegant: Use lean mirrors instead of bulky ones for the bathroom area with shades like bronze, cinnamon and dark chocolate. You can also buy modern mirrors with chrome finish.

Try mirrored furniture: You can buy mirrored furniture like a chest of drawers or coffee table with mirrors on it. If you have a small room, you can place them next to a wall and it will give an illusion of the floor extending beyond the wall.

Use the creatively in your hall: Take mirrors of different shapes and sizes and frame them. The frames can be ornate, antique, funky or anything that you like. Create a collection of these frames on the wall and they will look like artwork. Mirrors can also be used to lighten the look of your seating area if you have used vibrant paint on the wall.

Dos and Don'ts while using mirrors: Do use big mirrors in small rooms. This creates an illusion of adding more depth to the room, so go ahead and use a big mirror. Try having wall-to-wall mirrors Do use mirrors even in small spaces. Use a horizontal mirror in a narrow hall-way or staircase, to widen the space. Don't just use mirrors anywhere.

Don't hang a mirror to cover any empty wall space. Make sure the mirror reflects something good to look at and not a pile of unused things in the hall or a laundry basket in the bathroom.

Don't have too many mirrors in the kitchen since they need constant maintenance.

DÉCOR

There are simple ways to deck up your home for the festival without going overboard.

Sure, you have a beautiful home. But, there's always more room for a festive makeover. If your present décor theme is around functionality because of the huge traffic during the rest of the days, Diwali gives you just the reason to dress up your home differently. And it does not take that much effort to bring in the festive spirit. With just a week more to the vibrant festival, here are easy ways you can do up your home to go with the festive fervor.

COLORS ON WALL

If you haven't painted your home already and did not have the time for a complete overhaul, this is a quick way to bring in a splash of color.

HIGHLIGHT JUST ONE WALL

Choose a wall in the living room which draws most attention. Give this wall a dash of color or texture to make it the focal point. There is an entire range of textured paint available that gives you the freedom to experiment in creating your own textured focal wall at home. The new options in quick-dry-wall colours will make your task easy for the DIY trick. Once it is thoroughly dry, hang up your favorite artwork and fix an up-light or down-light to create the glow around the portrait.

If you have unseemly pillars or portions in the wall that need a quick layer of paint, do them up creatively
with floral textures in paisley motifs.

Gold paint is a top favorite for the festive season and having a textured wall in this hue can perk up your living area or bedroom elegantly.

DRESS UP WINDOWS

The weather in Vijayawada is just right to flaunt wispy pieces of chiffon or georgette to dress up your windows and doorways. Bring out those dressy, unused silk sarees and hang them up for effect. Try alternating the sarees to mix and contrast each other. For instance, cut a silk saree into two and hang up one of them as a window drape. Then put up a contrasting chiffon saree-drape. Do the same with the rest of the windows. The alternating silk-chiffon mix will exude an ethereal touch to the room and doorways.

String along some tiny brass bells or fairy lights across the rods to light up your windows for those vibrant Diwali parties.

CUSHION UP COMFORT

This is the best time to wrap your cushions in dressy fabrics for stunning effect. Make good use of leftover pieces of fabric. The unused silk saree blouse pieces are ideal for this. Or maybe, the embroidered pallu portion of the silk sarees. Stitch up two different pieces of squares or circles to make cushion covers, so you have two different colours and designs for the front and back. This way you can alternate them on various days of the festival.

Stitch one or glue on some sequins or tassels to the edges and you have the best dressed cushions in the house.

NATURAL CHARM

Wood flooring is an elegant way to furnish your floors and create an inviting ambience

With floor design becoming creative, more and more homeowners are focusing on making their flooring exciting to look at and walk on. From stone floors to marble flooring, tiles to carpets and even wooden, laminate flooring to the latest trend of leather flooring, new trends are taking over. With options aplenty, the aesthetics, durability, and budget help in the selection of flooring.

For natural warmth, elegance and traditional beauty, wood floors are considered to the best option. Very suitable for the entrance, living room, staircase and bedroom, wooden floors display a unique style that implies luxury. It gives an elegant appeal to homes like no other material can. Wood flooring often defines the size of the room, making a room looking bigger and lighter. Wood flooring is classified in three different forms. Hardwood flooring is the most common. It is economical when compared with other flooring choices. Made of fine, processed wood, hardwood flooring has a healthy sheen and has a thickness of 18 mm.

Wooden flooring is very easy to maintain. However, wood flooring is not recommended in humid areas like bathrooms and kitchens, and the use of adhesives should be avoided. Requiring extreme care and maintenance, the floor has to be wiped with a damp cloth and any spills should be avoided as it can damage the wood floor. Clean the floor thoroughly to remove dust and grit, then use gliding furniture pads underneath the piece to aid sliding it across the floor. Wooden flooring has taken over the concept of stone-flooring as it maintains the temperature of the room. There is a wide range of style available in wooden floors like strip, plank or parquet. Parquet floors, involve individual squares, laid down in a geometric pattern. It offers immense scope for your creativity and you can, in fact, make a striking styles statement with your wooden flooring by going in for inlays, insets and medallions. Artistic use of colours for inlays and insets and the right choice of medallions can work wonders for the floor and increase its beauty immensely.

STONE IN YOUR HOME DÉCOR

Stone imparts a timeless appeal to home interiors with its varied options

The use of stone in homes is an affordable option and is probably the oldest material that was used in home interiors as it cuts back on maintenance costs.

Exposed stone facades do not need plastering or painting and an initial extra cost of the material is compensated by the savings in the long run. They are available as large stone slabs or as tiles. In the earlier days, it was exclusively for functionality. Today stone is yet another cool accessory that SUITS any budget.

CREATING WARMTH

With earthy hues, it gives a warm look to the ambience. The textures, colours, and grains in the stones are totally unique, allowing it to age gracefully with time. Given its varied patterns, NATURAL stone can be used in multiple ways in an interior, resulting in a fabulous décor scheme.

Commonly used NTURAL stones are marble and different types of granite. While these stones are essentially used for flooring, countertops, and tabletops, they can also be used as cladding for walls. Semiprecious stones such as onyx and mother-of PEARL are used essentially as a highlight along with natural stones.

HARDY MATERIAL

Granite is a form of igneous rock and is ideal because of its hardness and non-porous nature and is available in a wide variety of colours. Black, red or brown families are harder than their lighter colored counterparts.

Granite these days is also available with a wide variety of textured finishes called antique finish, leather finish etc, where it is rendered rough by first flaming and again brushed with a polishing stone for the desired effect. This finish is widely used where an antiskid surface is required in toilets and also for cladding since it gives a more rustic look as compared to a polished glossy surface.

WIDE VARIETY

Marble is a form of metamorphic rock and is also widely used. Like granite, marble is available in large slabs. Marble is not pre-polished and polishing is done after laying on site. Stone can also be used as cladding in the form of slabs and tiles. There is a wide variety of slate stone, limestone and sandstone available.

Stone thus offers an almost limitless palette to accentuate and define spaces with and is used extensively today along with other materials to other a rich textural spatial quality filled with warmth and earthy flavors. The natural texture of stone is much appreciated for certain accessories like water fountains or sculptures.

GLEAMING FURNITURE

Some accessories revel in the smooth, polished look of stone to be used in the living spaces and beyond. Stone furniture is generally assumed to be rustic and more for the exteriors.

However, there are beautifully hand-carved furniture that can occupy a prominent place I the house. Carved sandstone armchairs can be easily used for the interiors. DINING TABLES, benches and other novel items are available in marble that go well with a lavish décor.

APPEALING ACCESSORIES

Stone tiles, decorative stone walls, arches and columns can be suitably tailored to be made as contemporary or traditional as required. From delicate looking onyx to unpolished granite, these stones look good alongside contemporary fixtures.

Consider wall shelves and counters of stone that match the flooring. Stone objects of art and accessories can be selected keeping the overall color in mind. Accessories can include dinnerware, planters, and even LIGHT fixtures.

MAKE EARTHQUAKE-RESISTANT BUILDINGS

THE EARTH-QUAKE ENGINEERING RESEARCH CENTRE [EERC] OF IIT HYDERABAD HAS INITIATED A MOVE TO MAKE HELP BUILDERS UNDERSTAND SEISMIC-RESISTANT BUILDINGS.

India has witnessed fast paced urban development in the last few decades, increasing the demand for commercial and residential establishments. As far as Hyderabad is concerned, in a run to transform into a smart city as well as a global one, it has become choked with wobbly development. The fast pace of construction activities with no planning has led to haphazard growth of medium to high-rise buildings, causing serious threat to life and property. Hence, there is a need to understand the risk from earth-quakes and take corrective measures.

The disaster planning such as mitigation and preparedness can enhance post-disaster responses such as emergency rescue and rehabilitation. Also, this reduces loss of life and suffering to a great extent. The main reason for causalities in a seismic catastrophe is collapse of buildings. To ensure construction of safe buildings, several issues need to be addressed. It is with this aim that IIIT Hyderabad has established Earthquake Engineering Research Centre (EERC) and has been conducting workshops to create awareness on earthquake-resistant structure designs.

In the IS 1893 code which deals with earthquake resistant designs for structures, the county has been mapped into five seismic zones based on seismic intensity. IS 1893 was first published in 1962 as Recommendations for Earthquake Resistant Design Structures and has been revised five times. In thc rcccnt rcvision, significant changes have been included: Design spectra are defined for a natural period up to 6 s and are specified for all buildings, temporary structures have been brought under the purview of the standard, additional clarity has been brought in on how to handle different types of irregularities in the structural system, effect of masonry infill walls has been included in the design of frame buildings.

Around 56 per cent of the land area in India is earthquake-prone (with 12% of land area coming under Zone 5, 18 per cent under Zone 4, 26 per cent under Zone 3, and 44 per cent under Zone 2). It is crucial for India to develop strategies for risk assessment and mitigation as 82 per cent of the population occupies 56 per cent of the land area. Hyderabad may not experience direct earthquake effects but can experience indirect ones caused by long distance waves. Sometimes, even when the hazard is understood, lack of knowledge on earthquake resistant design and the flaws of reinforced concrete structures heightens the aftermath of an earthquake. Professionals involved in the construction industry should preempt the eventuality of an earthquake while planning for the safety of buildings.

Seismic risk assessment has three components namely hazard, exposure and vulnerability. Vulnerability assessment of buildings consists of three stages. These are Rapid Visual Survey (RVS), Preliminary Assessment, and Detailed Assessment. Expended energy-based damage assessment model which was developed at IIIT Hyderabad can be linked to the Rapid Visual Survey (RVS) score of a building for assessing seismic risk. In this, the damage can be categorized as: a) no damage (b) slight damage (c) moderate damage (d) severe damage and (e) complete collapse. IIIT Hyderabad has also developed graphs to indicate the type of damage a building will have, when subjected to a particular magnitude of earthquake. This will help both individual home owners to know the vulnerability and risk status of their building and the government to know the status of the built-up environment in a given region.

The construction of seismic-resistant buildings is not very expensive. While a normal building costs Rs. 10 lakhs, a seismic-resistant building will cost only around Rs. 50,000 more. Not only buildings under construction, even completed ones can be made seismic-resistant. The technology is called Seismic Evaluation and Retrofitting.

Most builders are not concerned about providing earthquake resistance for buildings and are under the delusion that it won't happen in their area. But it is high time people realized the havoc an earthquake can create and should be prepared enough to tackle it. Builders should hire competent structural engineers and should ensure that they follow building bylaws and code specifications. They should make sure that contractors also follow the safety norms.

By standard evaluation and assessment methods, seismic-resistant buildings can be identified. Periodic evaluation and continuous education program are essential. Awareness needs to be built among the public on the eventuality of an earth-quake. One should understand that cost-effective and locally available technologies can be used to effectively prevent the effects of earthquakes. Earthquakes disaster management should be made compulsory in schools and colleges rather than sidelining it as an optional subject. There should be a peer review system for evaluating special buildings such as tall towers and buildings with more than 20 floors. Licensing for engineers is another important factor, making sure they are capable of providing disaster safety.

ENHANCE DÉCOR WITH WHITE HUES

Use white as a dominant color I your home spaces for an elegant look.

Color white was once considered dull and bring Today it is extensively used in home interiors to make the room look bigger and brighter. According to interior designers' vibrancy and character need not be brought in only by a splash of bright shades. Even a predominantly white interior with a dash of color appearing as highlights can look almost pristine, royal, lending elegance and charm to spaces.

The trick in white is to make sure you have a variety of tones, shades, and textures which gives a room enough variety so that it looks elegant, allowing you to make a statement about yourself. For instance, in a wholly white living room, a single color, say, aqua blue, appearing as an accent on cushions, or in curtains and as accessories can create a beach-like-vibe, providing color and character. Thus, motifs in squad blue could feature in soft furnishings, in photo frames, painted shells, even a soft floor rug, serving as a stunning contrast against the stark white background.

White in furniture goes well when the walls or at least one wall has a dark shade. The white here needs to contrast with an accent wall or rich colored floor rug. White objects of décor can be lit using colored lighting for effect. Usually, statuettes in white at the corners are highlighted using spot lighting in soft, light colours.

Interesting contrast can be created with the use of wood in and around the room, such as a top tables and paneled drawers. For a modern look, one can use glass center tables or crystal chandeliers Along with these, accessories like candle stands, vases in white, textured rugs or carpets in a neutral color go well. Ideally, a mix of different textures-leather, linen, fur, marbles should help create interest in a flat theme. Black, red, hot pink and orange are all colours that add interest and fill the otherwise plain interiors. Warm colours like chocolate help give an airy feel to the room. The latest trend is to mix white with metallic colours. Overdoing it, however, could be potentially disastrous, so it is advisable to use it sparingly. Use a well-paper in silver or gold in interesting patterns. A few wise choices while deciding on the materials, and viola, you have got the essentials to a stunning white décor. Try flushing in some natural shades in flooring window treatments or furnishing. Wood, cane and bamboo offer a number of options that would look stunning against a white back-drop. For a more contemporary feel, try working around glass and white metal. These would not weigh down the overall arrangement, but would carry off the sleek look well.

====

GO MINIMALIST IN YOUR HOME DÉCOR

This light theme will make your living spaces functional and manageable.

The beauty of simplicity is what minimalism aims to achieve and that is what one should strive for. It is all about carving space to the fullest with as less material as possible. A living room with understated coffee tables, basic yet contemporary lamps, a floor rug and a single color for the walls that's what the space should display. Put away the heavy curtains and throw pillows and don't crowd the space with photos and curios.

But while planning for simplicity, don't neglect comfort. Your sofa or futon should be comfortable and stick to a single color. For instance, a white color scheme with dark brown wood furniture is a wonderful contrast and combination which highlights a space well. This color contrast is enhanced especially by low furniture. If you have light colored furnishings on the sofa, then ensure they are in plain light shades and offset them with a dark color rug without any designs.

Another good way to high light your living room is by using a wall. Use a wonderful work of art. You can also give it visual appeal by using subtle textures or colours, or by skewing it to give the space a unique dimension.

While opting furniture for living room selects the ones which is simple yet classy. Wooden furniture with low sitting arrangements or center tables which can double up as dining tables can be used to save space. opt for a few pieces with solid, subdued colours. Plain, modern furniture ideally in natural wooden colours can be selected. Smooth fabrics with furniture in leather upholstery can add to the minimal effect.

Avoid using too many floor and desk lamps to light the room as this takes up unnecessary space and adds to clutter. Use of mirrors can create a dramatic effect.

The idea behind every décor concept is to create spaces that are friendlier and inviting and minimalist décor is just the perfect way to create more breathing space in the house. For those in the ethnic mold, try a simple diwan with a few cushions thrown in. Be careful to choose colours that are not heavy on texture or pattern as these can serve as visual clutter. Try solids and subdued shades.

As for windows, you can choose to leave them bare to allow lighter inside, or drape them in solid-colored curtains that fall elegantly. The key is to keep it simple. Tassels, ornate fringes and flares are a put-off.

Similarly, be restrained with regard to wall-décor. Wall-art is a great way of adding beauty to the home but remember that a few, very striking objects enhance the home A profusion of wall hangings of every kind covering the walls on all sides is off-putting and testifies to the homeowner's lack of discrimination or desire for ostentation.

Choose lamps and shades that are styled in minimalism. They can be in plain ceramic, brass or wrought iron but not too ornate. Soft lighting heightens the outlay and arrangement, in addition to offering subtle highlights to the few wall hangings you may choose to do up your wall with.

=====

SEASON FRESH KITCHEN DÉCOR

Spice up your kitchen with these style tips for a fresh new look for the season.

The kitchen is the most overlooked part of the house when it comes to decorating. While it is the most appliance-appointed and furnished room, the décor aspect is ignored in this high traffic food zone. This new season, infuses your kitchen with some fresh fabrics and furnishing to ring in zing.

DRAPED IN DELIGHT

Recreate those breezy English countryside kitchens in cheery hues of sunshiny yellows and grassy greens. Have pastel shades for the walls to make your task easier.

Begin with the windows. Instead of top to down curtains, go in for those that cover only half the windows. Choose fabrics in cotton in shades of canary yellow, sky blue or light green, with delicate paisley prints or floral ones with daisies on them. Tie them up in quaint styled bows in matching fabric or perhaps, curtain rings in wood with country themed miniatures painted on them. You can pick these up in any of the flea markets or at the malls.

For a change, put up matching door curtains too. These can be tied up elegantly with matching bows or ribbons during shifting of cookware to the dining area during meal time.

SEATING IN STYLE

If you do not already have a small seating arrangement in your kitchen, it's time to introduce it in a corner. Small kitchen? It's easier to do it up I a jiffy.

Fashion out a cozy kitchen nook for a quick cup of coffee or a tiny snack on the move Add a couple of moodas or those nifty ottomans to a corner or any empty wall in the kitchen. Place a cushion with a decorative cover atop the ottoman and you have a corner of your own.

You could also move in an unused chair against a wall. Stitch up a chair back in colourful material and add a matching cover to the cushion on the chair. Having vibrant mirror work on the cushions and ottoman tops can pep up your kitchen in no time to add that dash of festive décor.

LIVEN UP THE LINEN

Give your table and kitchen linen a huge and happy maker over this new season. You'll always find ways to make use of your old linen any way, as table wipes or even doormats.

Linen has a fresh touch to it that can uplift your kitchen in a jiffy. Use them as tale runners or tea towels; they'll never lose their style quotient.

Table runners in linen or any fabric that breathes work best in the kitchen. Cottons and linen have an airy look that enlivens the cook nook. In their chosen hues of yellows green or orange, hang them up in a row in their tea tower avatars. This on-the-spot festoon arrangement can look vivacious in the kitchen.

PRETTY POUCHES FOR POTPOURRI

To spread the scent of freshness all around your favorite room, you could be giving scented soaps a reincarnation. First, fashion out tiny colourful pouches or bags from leftover pieces of colored fabric from your kitchen décor exploits. Preferably, go in for satins or velvets for that regal allure. All you got to do is cut the fabric into handkerchief-sized squares. Now cut the soap into halves. Place each half onto the square cloth, gather up the sides of the cloth into a kind of flower, and tie it up with colored string or thin satin ribbons. Once you have about 10 such pouches, arrange them elegantly into a glass bowl.

You'll now have some fragrance wafting around your kitchen without the need for a room freshener spray.

Instead of soap, you could explore fragrant potpourri too. For a change, you could also fill the pouches with spices from the kitchen shelf, such as, cloves, star anise, cinnamon and cardamom.

=========

CREATE INVITING FRONT DOOR DECOR

Pep up the front door with enchanting décor to create that perfect welcome.

A STYLISH DOOR KNOCKER MAKES FOR AN ELEGANT FRONT DOOR ND IMPRESSIVE ENTRANCE TRADITIONAL STYLE BRASS ONES WITH TEMPLE MOTIFS GIVE AN ALTOGETHER ORNATE LOOK.

First impressions of a home are created largely at front doors. The front door of your home is the first and the last thing that visitors will see when they visit you. You can make a good first impression on your guests and visitors by adding some beautiful yet inexpensive decorative touches to this area. While the most common trend is to blend the entrance door design with the theme of the interiors, other ways to recreate an interesting entrance is by supplementing an existing door with accessories around it.

You may choose not to be in sync with any theme in particular but simply select from various designs of arched to rectangular framed door or a latticed doorway. These décor pieces do not need to be elaborate or out of the ordinary. Many times, just one or two especially selected pieces that reflect your taste will go a long way in making a lasting impression.

By decorating the front doorway area and the door itself, you can lend a personal touch to your home and make your guests feel warm and welcome. The design can range from contemporary to traditional, ethnic or rustic, modern and trendy, to any eclectic style. Any which way, it should be one that is best suited to the decor theme of the architecture and interiors of the home.

Coming to the min door itself, select a color for the door, keeping color theme of your living room or lobby in mind. So, a teakwood door with brass fixtures and floral carvings on it will suit a home with an ethnic theme. Antique doors brought down from an old home will look good for a home full of antique furniture and accessories.

A home with a minimal, Zen-like space could do with a plain, sleek and light-colored wooden door. A rich, dark rose wood door is ideal for a home with plenty of leather furniture. Now-a-days, the two-door concept is popular especially from the security point of view. The inner door is usually wood and the outer door can be metal or wood plus net.

For exteriors done up in dark colours or in bricks or stonework, a light-colored wooden door is complementary. If the door has to be painted, then white is the safest and smartest option. One can always try out a different color apart from white. Shades of purple, yellow or even red will work wonders provided the exteriors can support it.

A stylish door knocker makes for an elegant front door and impressive entrance. Today, among the more popular designs are traditional-style brass ones with temple motifs that give an altogether ornate look. These are especially, preferred by those who have built their home in a traditional or rustic architectural style and also whose interiors sport a collection of Indian antiques. Lastly, lighting up this area is important. Pick from a plethora of choices – lanterns to lend an old-world charm, a paper lamp for an Oriental look, or a colored lamp-shade for an unusual look. There are plenty of options.

========

FOR STYLE IN STAIRCASES

A tastefully designed staircase can uplift your living room décor.

Designing a staircase actually gives you an opportunity to announce your individual style. Well-designed staircases can make a dramatic statement and add beauty and drama to interiors with ornate railings and balusters. These days striking designs are chose for staircases in high visibility area like the main lobby, living and dining room. One major criterion for the design is the space available. Factors like material and style have to first complement the space. Styles like cantilever, floating, and spiral are good options in small to fair amount of space.

A simple stairway can be enhanced by its balustrade or railing, newels, and balusters. Though the utility of a balustrade is to provide a support system while ascending stairs, its aesthetics can make it visually appealing, and enhance the look of the stairs. The balusters are a series of upright elements fixed along the steps. The handrail is fixed at the top end of the balusters, tying them together and providing a handhold while climbing the stairs. These three elements must be designed and combined judiciously to look pleasing and complement the interiors of which they are a part.

Today, with staircases going from rectangular and circular to elliptical and helical, balustrades are not far behind in evolving new age contours and design So, whether it is the banisters or the décor, the design on your stairs or along the staircase can be made to blend in with the general ambience of your home.

DIFFERENT STYLES

If simplicity is your style, you could even go for simple, straight balusters that are un-adorned.

These blend well and disappear against the wall. Natural simplicity can be obtained with wood and other natural materials.

You must also match your floor with your stairs. Classic staircases reflect the style of a particular era and have a solid architectural style, elaborate with large posts and complex bracket designs. For maximum visual appeal you could go for a classic staircase with heavy architectural details, ornate cost iron balusters and decorative capping and handrails. Spiral staircases create an architectural statement while at the same time fitting into tight spaces. You could give your room a softer look with a circular staircase, or a more natural look can be had with carved wood railings and balusters. Spiral designs can be great for getting to out of the way places like attics.

Stairs can also differ in terms of whether they are wide or narrow, steep or gradual. These factors are generally decided by first considering the dimensions of the rooms or spaces it is going to connect and the overall interior décor plan, so that the staircase-style goes well with them.

DÉCOR ALONG THE STAIRCASE

It is essential to have eye catching objects of décor along the stairway. You can place framed works of art paintings, metal murals, or cloth wall hangings along the staircase wall. You can vary the placement height so they are in proportion with the rising stairs so that the piece is always at shoulder or head height at every step. Wall mounted flowers are another way to brighten such a wall. You can also have niches along the wall, either built in or wood wall mounted ones, in which you can place miniature art specimens. However, none of these features should be very large or they will overwhelm the staircase.

BEDROOM SPACE-SAVING IDEAS FOR NEWLY-WED COUPLES

A wedding in the family means adjusting a new member into the house. The first step to this adjustment is creating the space. Constructing a new section of the house or turning a bachelor pad into a master bedroom for a growing family of a newlywed couple can be an unenviable task. Here are few helpful tips.

SPACE-SAING SOLUTIONS

The most common problem faced is: There is not enough space. Multiple requirements such as bed, lounge, study, storage, dressing and bathroom have to be accommodated in the same space. With proper planning, maximizing space utilization, dedicated function allotment, modular fittings t your advantage, and thinking smartly about interiors one can create another bedroom worth cohabiting.

- ❖ Use a queen size bed instead of a king size. Opt for wider side tables for better storage and circulation space.
- ❖ Accommodates a chest of drawers in the room. It can go a long way in saving storage space for shoes and bed linens.
- ❖ Take advantage of wall height and accommodate shelving systems. They can take load of books, decorative artworks and trinkets.

- Combine the wardrobe and bathroom. Excessive walls consume space and create more hassle than utility in small spaces.
- Use sliding shutters for wardrobes as they reduce the circulation space required and provide more spacious wardrobes that can be wider than usual.
- Glass shower enclosures are more practical, and save space and water than bath tubs. They can be as small as 30 square inches in size.
- Large mirrors in the bathroom can give the illusion of space and be highly utilitarian. Often, new couples find it hard to agree on a color as women naturally tend to opt for warm and earthy colours and men prefer cool colours on a palette of grey-black.

DRIVEWAY DÉCOR

Enhance the path to your home with appealing décor elements.

There are various other ways to enhance the appeal of your driveway. Since this is a long stretch, which one has to traverse to reach the home, decorating it attractively will create a lasting impression on the mind of visitors. There are a range of options to choose from.

GARDEN GREEN

Having a beautiful home with a garden to complement it is a gream of every nature lover as it enhances the beauty of a driveway which looks empty otherwise. You have the option of having a garden spanning your entire area except for the driveway or if you want to make it more interesting, you could have several paths cutting across your garden leaving the largest for the driveway.

Pathways can be designed to meander between the lawn, fountain and patio. The patterns created by the path add to the beauty of the garden and hence the material and sizes should suit the type and area it accesses. All the elements like plants, pathways, water bodies, lighting and accessories should be planned to enhance the external ambience. Plant options could range from huge trees like the Ficus to smaller shrubs. You could have combinations of both for an even spread. For a more colorful look, you could combine green plants with others like crotons to that the garden has a range of colours.

MATERIAL MOSAIC

Types of paving range from concrete, brick, packed mud, small-sized jelly, cobblestones, natural stones like granite and slate, and even wood. Each of these gives its own accents to foliage due to the inherent size, texture and color. A combination of paving and plants gives the garden its character. Bricks pathways are fad these days, which are a natural alternative to concrete pathways it is both easy to lay and exceptionally durable. They are available as special paving bricks which are comparatively very affordable. Brick looks attractive in formal and rustic settings. Using unusual textures or arresting outlines can create interest.

Paving stones are the ideal paving material for small and large gardens and patios, as they are exceptionally easy to lay. Pre-cast concrete pavers can be lad in patterns and offer a choice of colours. Terracotta looks good also but due to their smooth finish can be slippery when wet. Stone paving is very durable. They lend the beauty of their natural qualities to the surroundings. Using unusual textures or arresting outlines can create interest.

LIGHTING ELEMENTS

Apt lighting features in the garden are can put its beauty under the spotlight at night. The garden is amazingly lively as much during the night with the tall pole lanterns lighting up the pathway, little spotlights along the interesting elements, between pants and highlighting the feature walls. Yu may integrate the lighting by having lampposts of the same design as the windows or of wrought iron if you have window grilles and blend the lighting with the elevation

Small bits of green shrubs and bushes near the lamp posts add visual interest. Depending on the kind of look you want to give your exterior, you can decide on the kind of bulbs you want to go for. There are the usual yellow bulbs, CFL bulbs or LED lights.

WATER BODIES

Water bodies along the driveway will also enhance its visual appeal. Urns, fountains, real or faux cascades, ponds and pools add to the aesthetic of your garden. They can be used in settings other than the garden also Yu could even arrange a stone path to the fountain, or combine a huge central fountain with small statues around it or kept at random throughout the breadth of the garden.

========

BAR IN THE LIVING AREA

A bar at home is a trend in interior décor that is catching the catching the fancy of homeowners.

One of the most common home décors trends these days is to have a small bar in fact it has become a must-have element in most of the homes these days. Thanks to the wide variety of readymade bars readily available in the market, a majority of young home-owners ensure that they dedicate a space to this fancy piece of décor. Many see it as a great way t enjoy a relaxing drink with friends, in the privacy of their own home bar, away from the noise and jostling interiors of the crowded clubs.

A bar could be a perfect addition into your patio, family room, dining room, living room or even your kitchen. Generally, and traditionally, it is always the living room or the dining room that provides space for the bar. A mini bar can be placed preferably in any corner or unit of the living room. For those who do a lot of informal entertaining, the family room or a large kitchen can also be chosen for a bar. If you live in a flat and have an open kitchen, a part of the kitchen can be converted into a bar. For example, a breakfast counter with glass cabinets below to store bottles and glasses could double up as a bar in the evenings. Alternately, even a corner cabinet in the kitchen can be converted into a bar.

Some prefer the area below the staircase as a location if it is within the living or dining area. The space allocated for the bar must include comfortable movement for at least one person behind the counter and for a couple of bar stools in front. One can also go for a bar within a wall cabinet which can open up when required. It can have a dropdown top, which can be used as the bar mixing table. Below t a small cup board can house the alcohol and glasses.

If required a mini-fridge can also be designed to fit in there, so when required the console can be pulled out and it becomes the front bar and the back unit becomes the back bar. Apart from lighting the bar under the shelves, highlighting a specific bottle would add to the look of the bar. Ne can also install some fancy bar equipment like bottle holder, hanging glass holders, wine stand. If space permits, hanging lights an also be used over the counter.

The design and size of the bar unit will depend on your needs and availability of space. The simplest bar arrangement could mean a bar table with a couple of bar stools. If you have a certain theme for the decorate your bar along the lines of the rest of the house. Otherwise, if you like you can adorn your bar in a different way from these rest of your hoe to set it apart from the rest of your spaces.

The bar can vary from a conventional semi-circular counter in a corner to a large trolley on wheels to even a cabinet tucked under the staircase that can swing out when required. Another option is to extend the crockery display unit in the dining room to serve as a bar counter. This can be done by providing a countertop for placing bottles and mixing drinks.

One of the chief aspects of a home bar is seating. You always want everyone to be comfortable and relaxed. Bar stools at the bar are great, but you should also consider appending some extra seating away from the bar if space permits. Lounge chairs covered in fun zebra prints or bright colored fabrics can lend an eclectric feel to your room You will need bar stools to accompany your home bar.

==========

LUXURIOUS IN LEATHER

Lend sophistication to your home space with this versatile furnishing material.

Comfort seating and loads of aesthetic value are big plusses that go with a contemporary décor. Leather furniture and furnishings have always been synonymous with panache, style and sophistication and has an ability to lend itself to any style from contemporary to traditional, and adds to a soothing, classy and attractive interior space. It is the brown leather furniture that lends a luxurious, upscale look to any room. Neutral browns or beiges can coexist with a variety of complementary hues to create a tranquil, soothing environment, or they can ramp up the palette with vibrant energy.

Sofas, chairs, sectionals and desks are the common category of leather furniture.

Leather chairs not only add style to your interiors but also provide comfort and space optimization while sectionals are also a great way to get the look and a feel of lavishness. Leather desks are also available in multitude of styles and are known for their sophistication and elegance. Apart from sofas, living room furniture can use leather for hairs, bean bags, ottomans, reclining armchairs and even bar stools. Dining rooms chairs, side-boards and settings could be in leather. In bedrooms, leather upholstery looks elegant for seating as in other areas like study rooms, and family spaces

Leather beanbags look classic in every home, adding comfort and warmth to the house. They are amongst the most comfortable pieces of furniture. They blend well with living room decors and are lightweight for everyone's delight. They come in a wide assortment of textures, shapes, patterns and colours. Additionally, it is quite versatile and can be used in furniture, as wall claddings and for paneling purposes. Versatility ad low-maintenance is its major advantages. If not leather furniture, then go in for accessories such as leather-framed mirrors, picture frames, lamps, area rugs and window treatments. A great innovation is the use of leather for window blinds. When combined with silk, wooden paneling in the room, plush carpeting on the floor and judicious lighting the effect can be exciting.

Paneling of cabinet shutters with leather combined with other accessories like metal studs can be a great statement in a room. Table tops in leather panels, picture frames, floor cushion covers, and mats complement traditional as well as a modern décor. A very elegant usage of this material is I covering antique books, which can be displayed in cabinets

======

FLOORING CHOICES FOR A READY HOME

The existing flooring can be overlaid with easy-to-lay choices.

With the number of nuclear families and working couples on the rise, time is the most precious commodity. Most families are opting for homes which require minimal effort to customize living spaces, catering to individual taste but with n compromise on style A home that is ready for its occupants is what everybody prefers

these days given the short waiting tie to possess the home.

FLOORING OPTIONS

One dominant element in the décor which dictates style is the flooring Apart from defining aesthetic value, flooring selection must be appropriate to the activity type in a given space. Different spaces in a home have diverse requirements. In kitchen and washrooms, maintenance may be the deciding factor whereas in living, family and entertaining zones, look and feel play the key role.

Ready homes may have flooring already laid in as per project specifications. However, the ambience need not be limited to the preset scheme. There are many easy to lay options on existing flooring with a wide range of finished.

HARDWOOD AND ENGINEERED WOOD

Hardwood flooring essentially consists of different species of wood, laid either as planks or parquet. Hardwoods give a warm appearance and feel. They complement both classic and contemporary interiors. Solid wood must bc nailed to the existing floor. But hardwood flooring is susceptible to warping due to excess moisture and is one of the expensive options.

A more budget friendly option is Engineered wood, which has a top veneer of expensive wood backed by layers of cheaper wood. Engineered wood can be nailed or glued on existing floors

LAMINATED FLOORING

Laminate flooring consists of a plywood or medium density fiber board with a decorative plastic laminate. A durable inexpensive option, laminate flooring is a very popular choice due to the wide range of colours and patterns available, even resembling wood and tile finishes. It is stained resistant and easy to install over existing flooring.

NATURAL STONE OR TILES

Natural stone offers the flexibility of being cut to required size and shape. Finished stone is moisture resistant and offers variety in laying patterns and groove inlay details. Similar to hardwood, natural stones can be expensive. A good alternate is the use of tiles ceramic, porcelain or terracotta. Natural stone and tiles can be laid on existing flooring using adhesive.

CARPET OR VINYL

Typically classified as soft floorings, carpet and vinyl are easy to install as wall to wall coverings on existing floors. Both these options are available as rolls or tiles and come in a variety of styles. Though vinyl is easy to clean, carpets need regular vacuuming and are not stain and spill resistant.

Other select options include trendy materials such as molded acrylic, glass flooring and metal decking. Eco friendly flooring materials such as bamboo and cork are also gaining in reputation as existing options.

Irrespective of the choice of flooring, certain precautions need to be taken while selecting and installing flooring in ready to move in homes. While experimenting with different flooring options for different rooms, especially I continuous spaces, level differences use to varied thickness of materials should be considered. Precise jointing is very critical to a sophisticated décor. The color, pattern and texture of the flooring must complement the overall design theme. With the right choice of flooring, exclusivity and sophistication can be redefined in any home.

======

PERK UP ROOM CORNERS

Decorating room owners with unique objects makes them focal points that uplift the interiors.

Nowadays the trend is to have cozy corners within the home that act as focal points of décor – spots that at times also turn into daily bonding rituals like having a meal, or even unwinding by reading a book or even just sitting down together to discuss your day. To beautify these corners, you need accessories of different kinds and beautiful focal points.

CORNER TABLE

A corner table can enhance your cozy corner. As a stand-alone piece it can be placed near the doorway or in a corner. These days, in many homes with large living rooms, a corner table is also a focal point and object of décor. A table with inlay work, glass and chrome elements, or stainless steel and brass elements is sought-after for both its utility and décor value.

These days, many designers experiment with corner tables-giving the glass tops with teak legs, designing the legs to resemble elephants etc. Silver candle stands on a corner table make for striking décor. An antique lamp with a shade can be kept on one too. Wood, bronze or metal bowls with potpourri or fresh flowers in water look good on a antique corner table. A tale can also be an accessory in itself to add elegance to a room.

ROCKING CHAIR

Antique wooden rocking chair will set the corner up. Decide on whether you want to show off the natural wood or have it upholstered to protect it. If you have a corner space, then investing in a rocking chair that suits the décor of the home is worth it. You can spend a few hours in a rocking chair curled up with an enticing read or just watching TV. Accessorize it with satiny cushions and watch how it will turn your corner space into a focal point in your room.

VASES

Vases with antique value and historic appeal can add a touch of style to any décor. They can be placed on mantelpieces, on a dining table, or in corners for a spectacular look. Artistically-designed Oriental and porcelain vases with landscapes, forests, and other beautiful scenes on them could be used as decorative pieces to highlight corners and niches. These go well with indoor plants too as green plants stand out in earthen vases, creating a natural appeal. Dry arrangements also go well with such vases, creating a natural earthy ambience.

COLOURFUL LIGHTS

It left unattended; corners can be dark places. Change that by bringing in some light. A set of paper lanterns hanging from the ceiling at different heights, illuminating the figurine of a white plaster of Paris fairy or a wooden doll at night, will be a treat to the eyes. Avoid harsh, white light but stick to soft, warm lighting to transform cold to cozy. Large brass lams placed on the floor in the corner of the living area can also accentuate your space. Lighting them in the evenings can make your room look dramatic. Pedestal lamps with stained glass lamp shades bring more attention to your room corner.

GREENERY, MIRRORS

You can bring life into plain corners by creating niches using plants and pebbles Try creating a shallow pebble bed to seat potted indoor plants. Or else, consider a hanging garden for that spectacular splash of color If space and resources permit, you can even consider doing up a small water feature that would provide a relaxing atmosphere.

Draw natural light into a room's dark corners by hanging a few well-placed mirrors. Hang small mirrors on the wall adjacent to the north-facing window so it reflects the west-setting sun. Place it across from a source of light, a nice painting or a window with a beautiful view.

=======

DÉCOR WITH COFFEE TABLE

LET A TRENDY COFFEE TABLE BE THE FOCUS OF YOUR LIVING ROOM DÉCOR SCHEME.

Choosing the perfect coffee table will transform your living space. Coffee table, which is at the center of your conversation area in your living room is the glue that helps to hold the pieces together. It should be used to make the room look more organized in style and function so selecting the right one is very important and since it commands so much attention, go the extra mile and transform your coffee table from a simple oversized coaster to a work of art. With the changing styles and designs from traditional to contemporary, classic to modern, coffee tables have always been the focal point for living rooms.

It is not just a centerpiece for the room but sets the look of the room and makes it complete. It is a place where one would like to relax and enjoy reading assured that the coffee is within reach. A range of items like vases, urns, shields and trophies, books, key containers and ornate boxes to store Knick knacks liven up this unit. Coffee tables are the perfect place for seasonal decorating-try a bowl filled with pretty seashells in the summer, a cluster of mini pumpkins and a vase filled with bittersweet branches in the fall, a glass jar holding shiny ornaments in the winter, and an orchid and stack of gardening books in the spring.

Depending on a room's theme, budget, size, shape and materials such that it matches with the rest of the décor-choose the one for you. Many of the traditional ones are available in square, circular shapes with different styles and sizes. With people going in for customized furniture even stylish and trendy coffee tables are used to highlight the décor.

Combined use of wood, glass is used to make it look more an artwork. Also, wood with marble, semi-precious stones, marble tops along with carving. Swarovski crystal is also in. Though people opt for contemporary style of furniture there is fusion of classical and contemporary styles with more detailing to get a warm 'feeling' that goes along with the rest of the décor. If you have a clear acrylic or glass coffee table, don't forget to make use of the display space below If your coffee table isn't see-through, you can still get the same effect with a patterned rug. A glass or Lucite table makes a room look larger as you see through them. Materials in wood like solid American white oak and American black walnut veneers works well in any type of living room because they easily straddle both traditional and contemporary design. It is great designed and comfortable coffee table to impress your guests.

THE TREND IN COFFEE TABLES IS TO HAVE LOW LVEL ONES WITH SEATING ATTACHED THAT CAN BE TUCKED UNDER THE TABLE WHEN NOT IN USE. THE LOW SEATS CAN BE MOVED OUT DURING A COSY GET TOGETHER.

An oak table will complement any room I your home or apartment. One look at the elegant simplicity of this table and you can see why neo-classical style is still popular in furniture design. While somewhat formal in appearance, the clean, straight lines of this table would work just as well in a modern or contemporary room.

Some are available in the form of storage spaces with drawers and shelves underneath which are used to hold magazines and books. These days coffee tables can be easily adapted for multifunctional purposes. Coffee tables that offer storage underneath are great for small spaces. These tables can be designed with proportions that will allow stools to slide under the top for storage when not in us, freeing up the floor area that is at a premium. These units can also be designed with a lower shelf, drawer or stand to display artifacts, hold books or stack music paraphernalia.

Shelves, drawers and cubby holes really help keep the clutter of remotes and piles of books or magazines handy.

=====

THE SIMPLE HACKS FOR A CONTAINER GARDEN

Not everyone is able to make up with a green lawn or space to bloom in their gardening tactics. One can use fish pots, glass lamps, balls and bowls to create their own hanging gardens or reuse old furniture as garden desks. We suggest few simple hacks to bloom a container garden.

HANGING GARDEN:

Use fish pots, glass lamps, balls and bowls to create your own hanging gardens. This does not require much effort and can be put up anywhere. You just need to plant your favorite's and hang them in your balcony, patios and entrances. You can also go for sea-shells that can be decorated to give décor a completely new definition.

OLD FURNITURE TURNEED GARDEN DESK:

One who is having thought of replacing their furniture, must give old one a new green look. You can place plants in the chest of drawers as well as on top. You can place it both inside and outside your room. Decorating it with nostalgic articles and making your space look livelier and greener.

VERTICAL GARDEN:

The simple technique to get this look is to place planters in small pots, cans or glass jars and affix them on wall through a wooden frame that gibe a bright sunny sight to the plain, monotonous wall.

THEMATIC GARDEN:

One can go for single theme like by having planters of same clan. For example, in shape of birds, animals, letters, fruits and vegetables, or any other design that defines your personality or taste. It enhances your walls, balconies by just plant in small planters requiring less water to dwell.

======

MAKE AN IMPRESSION ON YOUR GUESTS

Comfort and elegance should dictate the décor and plan of the room dedicated to your guests.

During holidays and festive times of a house becomes a busy place with relatives and friends pouring in to holiday or celebrate together. Many guests visit you and carry away a lasting impression on your décor and interiors. Making guests eel welcome and comfortable at home is the mark of a good host.

To do that, the first thing you need is a comfortable guestroom It should be an interesting space that looks impressive and above all makes your guests feel at home. Besides being comfortable, guestroom furniture should be attractive as well as functional. It should be decorated I such a way that it can also be used as a rest or relaxation too at other times by the occupants of the home.

Depending on whether the interiors are modern or traditional, the bed stand, side tables seating area and cabinets can be designed with best material to ensure a sophisticated ambience. The space should exude a feeling of comfort and luxury-all décor items, artifacts, elements, fittings, fixtures, accessories and equipment must be top-of-line in terms of quality and look.

A traditional guest room includes clean and simple antique furniture that never go out of style. Many pieces of antique furniture that never go out of style. Many pieces of antique furniture can blend in with modern or contemporary décor, giving the room an overall rich and luxurious feel. Antiques like majestic recliner chairs, ornate chests, and tables and elaborate hand-carved patterns can find a place in guestrooms.

These days, contemporary furniture is easy to find and convenient to use. It is often the best choice for guestroom since it fits into all houses, and offers a wide variety of choices that are guaranteed to suit any taste, comfort and style and is also extremely functional.

You also have a wide choice when it comes to designs and colours. Sleek steel tables and chairs will give the guest room a sophisticated yet comfortable look. Low-rise beds and tall, slim closet speak of futuristic design. There are a number of themes you can build into your guest room. Most popular being ethnic-this theme can be both simple and colourful. Use a lot of silk for a luxurious look and feel. It gives an incredible feel to upholstery. You can also fashion out silk bedspreads and pillow cases from your silk sarees. The varied colours used on these will instantly brighten up your room.

For the walls, you can use colourful hand-woven wall hangings. Ethnic paintings are popular among art lovers around the world. Paintings in earthy colours will add to the ethnic theme.

While the furniture itself may not be very colourful, you can add color to the room with vibrant and comfortable cushions, pillows and carpets. It will give your guestroom a very cheerful look.

Have a small mirror in the guestroom. If space allows, a long dressing mirror is an even better addition.

Instead of harsh overhead lighting, try bedside sconces. Not only do they cast a soft, soothing light, guests won't have to negotiate their way back to bed in an unfamiliar space

=====

ELEGANT APPEAL WITH STONE

Semi-precious stones and metals as part of the décor can bring a regal touch to your interior spaces.

These days, many prefer to combine various traditional and modern elements of décor to achieve a regal appeal in their interior spaces. Some opt for modern furniture embellished with stones and inlaid with previous metals. In fact, embellishing your home with traditional décor rooted in luxurious jewelry-like elements can be exciting. Glimpses of bejeweled marvels are now seen in contemporary furniture collections often using modern elements such as crystal, lapis lazuli and metals in their embellishments.

Sofa sets and single sofa chairs look elegant with crystal studs in the plush upholstery. The furniture style has been complementing the luxe look. Customize your living room furniture of leather and wood to include glimpses of gold in the frames too. Often, mother of pearl too goes into the design to make a pretty contrast to the precious metal.

ROM INLAY WORK TO SUBTLE ELEMENTS OF STONE AND MTAL THERE IS A WIDE VARIETY TO CHOOSE FROM. YOU CAN GET YOUR FURNITURE TO ADD TO THE AMBIENCE AND INTERIOR DÉCOR OF YOUR LIVING SPACES.

Subtle gold elements can be infused into any furniture style today. Cushions lined with tiny winkling crystals can look magical. Marble is versatile and can be used for table bases. The inlaid coffee tables come in artistic designs and luminous colours, and can be mounted on a variety of bases that vary from wrought iron to marble. Due to its beauty and elegance, marble is the preferred choice for the most precious end or coffee tables. Inlaid tables can have varied figures and forms – square, rectangular, circular or elliptic. Marble inlay panels in rectangular shapes bring splendor to a home.

Another possibility is marble inlay paintings on the walls-from ancient inscriptions to a village scene there is plenty to choose from. You can also have inlay work done on marble plates.

Subtle gold elements can be infused into any furniture style today. Have a whole colourful medley of semi-precious stones to enhance your lamp-shades, table tops, jewelry boxes and dressing tables, in a flurry of designs. Statues or figures in stone such as black onyx, coral or aquamarine are also trendy. Statues adorned with hunky, beaded gems bearing intricate designs, dozens of oversized, gem-studded bracelets; earrings that resemble clusters of dangling grapes; thick anklets; and jewel-encrusted nose rings are also popular. Gemstone inlay work on other stones or other metals can be stunning.

If you have a legacy of furniture heirlooms in your possession, give them a precious makeover by embossing them with jewels. It will make them all the more priceless. Jeweled mirrors and chests bring glitter to any space. In the hallway, they would make a grand entrance, in the living room, a great conversation piece. If you own bejeweled collections and precious heirlooms in gold and silver, highlight these in a corner of your living room, dining room or the library.

You can even opt for floor medallions with inlay work-they look majestic in the central hall or lobby. A scheme which incorporates both traditional as well as contemporary elements can have jaali work along with wall and floor inlays. Floor medallions can have simple designs with thin, plain borders.

a traditional décor scheme, artifacts in bronze and wood will lend a stately touch. Small niches where the inlay work is prominent, on the wall as well as on the floor, can hold figurines. A chandelier above a large floor medallion with elaborate detailing adds drama.

REVEL IN THE FRESH FRAGRANCE OF FLOWERS

The aroma of fresh flowers can uplift one's mood and create a cheerful ambience.

If ever there was a right time to get flowers back into décor, it's now. But did you know that there is a flower for every type of room in your home? Interior décor experts tell you that a flower can either make or mar the look of the room depending on its size and shape. Here's all you need to know.

RED ALERT

For a wide formal drawing room, it is always advisable to use red flowers such as roses or carnations. The reds enhance the beauty of the room, making it look spacious and royal at the same time. As for the arrangement – take a pick from a traditional urli or a longish vase.

GET THE PASTELS OUT

If you have a narrow formal drawing space, use pasel coloured flosers such as white lilies or light pink or chides to give a soothing touch to your living space.

BAMBOO BASICS

In addition to their decorative appeal, bambook also helps purify the air. Since the plant has the tendency to grow relatively tall, you should find a place for it on the floor itself. There are ways to control the growth of these plants and create all kinds of quirky designs. Make one according to the width of the room.

LOTUS ARRANGEMENT

For a big balcony space or a terrace garden, grow a lotus stem in a concrete urli or a sturdy, waterproof pot in a sunny posttt. The arrangement can be placed in the center or on top of a traditional table.

WHY GO FOR FABRIC FLOWERS?

Silk, satin and tissue flowers are colourful and washable making them a convenient option for your home.

Can be made at home by following a few simple instructions.

Lasts a long time – they come in a huge variety.
Arrange with other potpourri and colourful pebbles.

Make your bedroom a comfortable sanctuary

With some cool ideas, you can turn your bedroom into a luxury zone for precious sleep.

Is your bedroom helping you sleep better? As sleep becomes a much talked about wellness concept, bedrooms are turning into sleep sanctuaries. Sleep entrepreneurs and designers are getting innovative. From swinging beds to weight blankets – new furniture forms are being invented. For a good night's sleep, it's essential that your bedroom has little furniture and is clutter – free.

The bedroom is the most luxurious and personal space in any home. Beautify your room with posh cushions and rugs. They'll help muffle noise, in addition to looking warm and cozy. An important element for a good sleep is lights. Add table or floor lamps in your bedroom to free yourself from stress and worries. Create your personal sleep sanctuary.

GET A GOOD MATTRESS

A good mattress is an investment in health. If your mattress is too hard or soft or lumpy, it will keep you up all night. Most people wake up with neck and muscle pain because of a bad mattress. It's a good idea to try out your mattress before paying the whole amount.

MAKE YUR ROOM DARK

We need a dark, calm environment to sleep better. Most modern devices like digital clocks, televisions, phones, light up your bedroom and interrupt sleep. Remove them from your bedroom or put them away.

CUT OFF THE NOISE

Is your fan and AC noisy? Is your room too hot or cold? Install a good thermostat to control the air temperature in your room.

HEALIING SOUNDS

Play soft music – like soft wind chimes, birds chirping or even the sound of falling water in your bedroom You could keep a small fountain near your bedside.

GET GREENS INDOORS

Keep a small plant on your dressing table or bedside to purify the air. This also increases humidity and helps you breathe better.

======

FOR REFRESHING MONSOON DÉCOR

GIVE YOUR HOME A VIBRANT MONSOON MAKEOVER WITH FOCUS ON COLOURFUL, WARM FABRIC AND AMPLE WHITE LGHTING FIXTURES.

Trends change with seasons and so does the style of your home décor. Changing furniture may be taxing but furnishings can change with times. With options increasing and quality of fabrics improving, one should look out for practical and long-lasting fabrics. With the wide variety of fabrics, textures and colours available in the market, the choice can be very confusing. With the onset of the monsoon, there is more humidity and moisture in the air. The most convenient way to add warmth is to introduce a lot of fabric. Monsoon calls for light fabrics and therefore fabrics with synthetic blends are always a good idea.

SEATIN BASICS

Spruce up your sofa chairs with extra cushions that are soft and springy. Change your cushion covers with brighter shades of yellows, oranges, reds, blues and greens. Choose colours that complement each other and give your home a cheerful aura. Minimize the use of pastels, whites and beiges. Like your clothes, you can dress up your home in light fabrics that are quick to wash and dry.

Add a couple of bean-bags for extra seating. You can also consider throwing a thin mattress decorated nicely with bolsters and covered with a thick quilted bedspread, beside a large window. This can be a beautiful spot to watch the rain and unwind while sipping a hot cup of tea. Keep curtains to a bare minimum. More fabric will attract more fungus. For your bed, use light-weight fabrics for sheeting as they are easy to wash and dry. Heavy-weight fabrics do not dry quickly resulting in a constant dampness and a foul smell. If you have a small room then choose furniture that is compact with cushions in neutral shades. This makes your furniture look ore spacious.

DRAPES AND RUGS

Replace sheer curtains and blinds with thick drapes. The best option is to use right colored synthetic curtains to not only brighten up the day but also avoid the need for dampness in the air. Also let the fabric be colourful to nullify the gloominess of the weather. Ensure you avoid dark colours and go for cheery yellows, bright pinks and glowing lilac shades.

The floor, especially if it is marble, tends to be very cold throughout the day. This can cause a lot of discomfort and health hazards too. The best way is to invest in a couple of carets, either woolen or synthetic. Lay them on the bare floor. They add beautify and add a touch of lavishness to the decor. Alternatively, you can also use woolen throws or silk rugs for a traditional appeal.

Cover your dining table with a thick, handloom table cloth and add another one that is easy to wash and dry. Use thick cotton tablemats and also slip-on back covers for the chairs for warmth.

LIGHT UP RIGHT

Rainy days are often dark and gloomy with bare minimum natural sunlight being available. Change some of the white lights in your home to yellow lights for soft mood lighting that can lift your spirit without being too jarring. Trendy table lamps and floor lamps in Indian motifs are perfect, as are modern, geometric shapes or matte steel lighting fixtures coupled with glass for a contemporary décor style. And don't forget candle décor for further mood lighting Aromatic candles will infuse your living spaces with fresh scents that keep out the musty odours of the rains. Also mirrors hung on the wall can also help brighten p a dreary dark space. Bright colored paint can also help to bounce light around a room.

====

A SEPARATE STUDY AREA AT HOME

Carve out distinct study space out of the living room or bedroom for privacy.

If you have kids at home then it's important that you create a separate cheerful study corner to encourage their academic pursuits.

They should be provided with that calm and silent space where they can study without any distraction. However, due to space constraints in most homes, this becomes impossible. This said, the next best thing is to have a small study corner in a bigger room- whether in the living room, bedroom, or dining rooms.

IDENTIFY BEST LOCATION

Whichever area you identify, make sure your child uses this space to study on a regular basis. Do not change the location every day or every week. First, clean and clear the area and demarcate the space. This is merely to assign a separate space for studies. Even the smallest space can be turned into a library-cum-study, when that space is used imaginatively.

FURNITURE BASICS

Though it is nice to have good looking pieces of furniture, ideally, the study table should be more utilitarian So, it is better to have a simple, sturdy structure with non-distracting design elements The table should have enough storage space with drawers and shelves to sore books and other items.

The surface should be anti-glare and should have adequate space to allow the child to keep books handy. It is best to have storage space for work material above the study desk as well.

It's important to pick the right chair as well. The chair should be comfortable and ergonomically designed. You do not want your child to stoop or develop back problems early in life. Ideally, install a table, which is large enough to accommodate a computer also.
Books can be stored in open or closed bookshelves. You can have standalone cupboards or wall shelves.

RIGHT LIGHTING

Lighting in this area is very important. Have a desk lamp or a bright light just above and in front of the desk area so it does not create shadows over their books.

COOL COLOURS

Select calm and colours that create a soothing atmosphere for the study. Warm, bright colours such as orange, light green; beige, yellow or peach are suitable as they will lend a cheerful look to the space.

DÉCOR AND INERIOR DESIGN TIPS

Mantri PragadaMarkandeyulu, Litt.D.,
Poet, Novelist, Song and Story Writer
B. Com, DBM, PGDCA, DCP.,
Email: mrkndyl@gmail.copm

CPSIA information can be obtained
at www.ICGtesting.com
Printed in the USA
LVHW100346011222
734242LV00009B/159

9 789356 754294